THE
EDUCATION
OF KOKO

THE EDUCATION OF KOKO

Francine Patterson & Eugene Linden

PHOTOGRAPHS BY RONALD H. COHN

An Owl Book

HOLT, RINEHART AND WINSTON · NEW YORK

QL 737
P96
P37

Library of Congress Cataloging in Publication Data
Patterson, Francine.
The education of Koko.
Bibliography: p.
Includes index.
1. Gorillas—Psychology. 2. Human-animal
communication. 3. Sign language. I. Linden,
Eugene. II. Title.
QL737.P96P37 599.88'460451 81-1325

ISBN: 0-03-046101-4
ISBN: 0-03-063551-9 (An Owl book) (pbk.)

First published in hardcover by Holt, Rinehart and Winston in 1981

First Owl Book Edition—1983

Designer: Helene Berinsky
Printed in the United States of America
1 3 5 7 9 10 8 6 4 2

ISBN 0-03-046101-4 HARDBOUND
ISBN 0-03-063551-9 PAPERBACK

To Koko and Michael

"Fine animal gorilla."
—KOKO

CONTENTS

ix

Photograph sections follow pages 45 and 151.

PREFACE

Koko is a ten-year-old, female lowland gorilla. She is the first of her species to have acquired a human language. This is the story of Project Koko, the longest ongoing study of the language abilities of an ape yet undertaken. The project was initiated by Dr. Francine Patterson in 1972 and is still continuing today.

The Education of Koko is the cooperative effort of Eugene Linden and Francine (Penny) Patterson. Eugene Linden has written extensively about the various language experiments with the great apes, and it is his feeling that Project Koko has achieved the most extraordinary results of any of the language experiments with animals. As this book will be dealing primarily with Dr. Patterson's research and the events that have marked Project Koko, the authors have decided to use her voice to present the details of her work. The interpretation of these details reflects the consensus of both authors.

ACKNOWLEDGMENTS

I owe an enormous debt of gratitude to Dr. Ronald H. Cohn for his continuing support, advice, and help. Throughout the entire project, he has shared the responsibilities, hardships, and problems but has rarely received recognition for his efforts. Dr. Cohn has done an outstanding job of photographically documenting Koko's achievements during the past nine years.

My sincere thanks also to Barbara Fallon Hiller, Koko's oldest and dearest friend, who has contributed to the success of the project in innumerable ways, generously donating both time and money. Barbara's comments on the manuscript in its various phases were invaluable.

There are many, many others who have provided invaluable assistance along the way, and I wish to extend my thanks to each, particularly the volunteers who worked on Project Koko. Here I would like to single out just a few deserving special mention: Ronald Reuther, who as director of the San Francisco Zoo granted me permission to begin the project with little Hanabi-Ko; Edward Fitzsimmons, attorney at law, who helped establish the Gorilla Foundation and for years has generously donated his time and advice; my late mother, Frances Spano Patterson, whose loving spirit provides inspiration; my father, Professor C. H. Patterson, who has always given me encouragement and support; Dr. Jane Goodall, who has aided the project at critical times over the years; Mrs. Clay West Head, who provided funds to help secure Koko; and all the members of the Gorilla Foundation, whose kind letters and generous contributions have heartened and sustained us.

I would also like to express my deep appreciation to the National Geographic Society for awarding grants in support of my project from 1976 to the present, and to the Djerassi, Favrot, and William Penn Foundations for providing timely grants-in-aid for specific needs. And for recent and most wonderful contributions for improving the gorillas' habitat, I extend my thanks and heartfelt gratitude to Mrs. Henry Doelger.

Finally, I wish to make a special tribute of thanks, praise, and love to Koko and Michael.

—FRANCINE PATTERSON

I would like to thank all the Project Koko workers and volunteers who helped me during the researching and writing of this book, particularly Barbara Hiller, Ann Southcombe, Kris Hanson, Barbara Weller, Maureen Sheehan, and Malka Kopell. And, for their stimulating conversations about various issues raised by Project Koko, I would like to thank Lee Blaine and the other logicians and scholars I encountered at Stanford University during my periodic trips out west.

Jim Wilcox and Iver Kern read sections of the manuscript at different stages of completion, and had many excellent suggestions.

I owe my greatest debt to our editor, Natalie Chapman. Throughout the writing of the book, Natalie showed prodigious patience, tireless attention to detail, and deft editorial judgment.

—EUGENE LINDEN

PART I

Getting Started

1

Conversations with a Gorilla

When I began teaching Koko American Sign Language nine years ago, I had no idea how far she would progress with it. There was little reason for me to assume that a gorilla could learn to use language to rhyme, lie, joke, express her emotions, or describe her world.

Nor could I have anticipated that the intense controversy ape-language experiments generated within the behavioral sciences a decade ago would still be continuing today. During the past few years the idea that any nonhuman can acquire language has been denounced with renewed vigor, and yet ironically it is also within this time that Koko has begun to demonstrate her most remarkable abilities.

Just how far those abilities extend is difficult to answer. Take one simple example. A visitor recently stopped by to see Koko. On greeting the 180-pound gorilla, the visitor pointed to her and then made a small circle with her open hand in the air in front of her own face, signing *You're pretty*. Koko digested this comment for a moment and then stroked her finger across her nose; her reply meant *false* or *fake*.*

Was Koko's response an indication of modesty, or a comment on her visitor's sincerity? Was it a random gesture carrying no significance? Was she simply imitating someone else's previous response to the same compliment? To prove what Koko

---

*Throughout the book all signed words (i.e., those made in American Sign Language) will be indicated by *italics*.

meant—or that she had any feelings about her looks at all—is a maddening proposition. It means establishing that Koko in fact made the sign cited, that she knew what she meant, and that her behavior was intentional, not imitative or cued.

That is the job of this book—to show how Koko learned language, and *that* Koko learned language; and to look at what a gorilla does with human language.

Why does anyone care whether or not an animal can learn language? This issue has intrigued humankind from Plato and Descartes to contemporary scientists and thinkers, for thousands of years. But its importance was perhaps best expressed recently by Walker Percy:

> Where does one start with a theory of man if the theory of man as an organism in an environment doesn't work and all the attributes of man which were accepted in the old modern age are now called into question: his soul, mind, freedom, will, Godlikeness?
>
> There is only one place to start: the place where man's singularity is there for all to see and cannot be called into question, even in a new age in which everything else is in dispute.
>
> That singularity is language.
>
> Why is it that men speak and animals don't?
>
> What does it entail to be a speaking creature, that is, a creature who names things and utters sentences about things which other creatures understand and misunderstand? . . .
>
> Why are there not some "higher" animals which have acquired a primitive language?
>
> Why are there not some "lower" men who speak a crude, primitive language? . . .
>
> Why is there such a gap between nonspeaking animals and speaking man, when there is no other such gap in nature?
>
> Is it possible that a theory of man is nothing more nor less than a theory of the speaking creature?

When Walker Percy wrote these words in 1954 in *The Message in the Bottle,* he could speak with confidence—and find unanimous support from scientists—about the fact that only man might learn language. According to the traditional wisdom of the behavioral sciences, animals can only signal. Their communication consists of a preprogrammed series of instinctive reactions to the immediate demands, fears, and pleasures of their lives. In the 1960s, however, a series of experiments involving two-way communication with apes began to erode that traditional wisdom.

Language-using apes have not only destroyed our confidence about the uniqueness of language—and therefore of man—but have also exposed uncertainty in the scientific world about what exactly "language" is. And the experiments have raised the question of what the apes are doing when they communicate with their human mentors. Are they in fact using language, or are they merely interpreting nonlinguistic signals unconsciously given by the experimenters? In short, have apes learned language or have they learned a circus trick?

If, as we hope to show, claims that we can talk with the animals are legitimate, then what they have to tell us far outstrips what we might imagine. In the nine years during which she has been taught American Sign Language, Koko has learned not only a large number of words, but also a great deal about language. It has become an integral part of her daily life. The language Koko uses, American Sign Language, or Ameslan as it is called by the deaf for whom it is a primary mode of communication, is the fourth most commonly used language in the United States. It is not English. It is a gestural language, and there are marked contrasts between the way a statement is made in English and the way it is made in sign language. For instance, it takes on an average about twice as long to complete a word in gesture as it does to say an equivalent word in English. This constraint places a premium on economy of expression. (Thus, the written translation of statements made in sign language has a stilted, telegraphic quality.)

Koko's conversation has changed dramatically through the years. At age three, Koko was manifestly an infant. She showed a great deal of dependence, a lot of brattiness, and relatively little signing in general. Many of her attempts at signs were unclear or inappropriate. A high percentage of her statements during this early period were requests for some form of sustenance or stimulation (tickling, chasing, swinging—these were very frequent requests). Indeed, a reading of the records might give the misleading impression that Koko was living on the edge of starvation and getting by precariously on handouts: *Pour that hurry drink hurry... me me eat... you me cookie me me... gimme drink thirsty,* and so on.

By age six, she was exhibiting her own ideas about language and the uses to which it might be put—such as expressing her increasing independence. One day when Koko was six I came in at 6:00 P.M. to put her to bed and relieve Cathy Ransom, one of my deaf assistants. Before leaving, Cathy pointed to the notebook in which all of Koko's utterances are logged. There I found Cathy's transcription of an "argument" she and Koko had just had in sign language. The dispute had begun when Cathy showed Koko a poster picture of Koko that had been used during a fund-raising benefit. Cathy had signed to Koko, *What's this?* by drawing her index finger across her palm and then pointing to the picture of Koko.

Gorilla, signed Koko.

Who gorilla? asked Cathy, pursuing the conversational line in typical fashion.

Bird, responded Koko.

You bird? asked Cathy, not about to let Koko reduce the session to chaos.

You, countered Koko, who by this age was frequently using the word *bird* as an insult.

Not me, you bird, retorted Cathy.

Me gorilla, Koko answered.

Who bird? asked Cathy.

You nut, replied Koko, resorting to another of her insults. (Koko switches *bird* and *nut* from descriptive to pejorative terms by changing the position in which the sign is made from the front to the side of her face.)

After a little more name-calling Koko gave up the battle, signed, *Darn me good,* and walked away signing *Bad.*

Cathy and Koko's argument illustrates one of the principal lessons of Project Koko, which is that in being "bad," Koko can be very, very good. Throughout the nine years of the project, Koko has been driven to her most creative uses of language through her obstinate refusal to submit meekly to dull routine. Indeed, the most telling proof that Koko understands the language she is using is the way she adapts it to express her impatience and other feelings.

Today, at ten, Koko is somewhat less mischievous, and much more verbal, than she was at three. In Koko's conversation today we see her ability to "build up" complex ideas through a series of short statements. How Koko does this, and the thoughts she expresses this way, is what this book is about.

2

Getting Started

Project Koko began in July 1972, the day after I received permission from the San Francisco Zoo to attempt to teach Ameslan to an infant gorilla. I had had my eye on this gorilla for nine months. In fact, I had begun planning Project Koko the day I first saw little Hanabi-Ko, or Koko, as she was nicknamed. And months before I first saw Koko, I had decided that I would devote my graduate education to the study of the language abilities of animals.

I was inspired by a lecture delivered at Stanford University by Allan and Beatrice Gardner, the comparative psychologists who first succeeded in teaching language to a great ape. This was in September 1971, five years after the Gardners had begun their work with Washoe, a chimpanzee, and ten months before I was to begin working with Koko. I had read some material on the Gardners' research, and wanted to hear them describe their methods and their accomplishments and see the films of Washoe conversing in sign language with her human companions.

As the Gardners described how they got the idea to teach sign language, their search by trial and error for a proper teaching method, the elaborate controls they developed to ensure that their data were reliable, and finally Washoe's willing response to their efforts to teach her language, I felt increasing excitement. Clearly there might be untapped language abilities in other animals as well. Although the Gardners delivered their lecture soberly, I felt that I was hearing about something from the realm of myth or fable: Animals were capable of telling us about themselves if one knew the proper way to ask them.

This lecture gave focus to my lifelong interest in animals. I started planning to try to find an ape and the funding that would permit me to pursue research along similar lines, and I enrolled in a course in American Sign Language. My inclination was to work with chimps because they were noted for their tractable, gregarious nature. At first I did not entertain the idea that it might be possible to try to teach language to a gorilla. But I would have leapt at the chance to work with any great ape.

I did have some background working with primates. I had entered the doctoral program in psychology at Stanford in the fall of 1970, after receiving a B.A. in psychology from the University of Illinois and traveling west with Ronald Cohn, a molecular biologist and close companion who has devoted all his free time to Project Koko since its inception. My interest in psychology came from my father, who is a professor emeritus in educational psychology at the University of Illinois and has published several books on counseling and psychotherapy. For me, however, graduate work in psychology was attractive because it would permit me to work with animals.

To the nonresearcher, the idea of a behavioral scientist "working with animals" often conjures up an image of the horrors of vivisection. This was not what I had in mind. I count myself among a "new breed" of behavioral scientists who would rather observe an animal than take it apart. We are more interested in understanding animals in their own right than in seeing how they might be used to understand and cure human problems. Indeed, the most delightful aspect of my work with Koko is that language allows us to see the complexities and subtleties of the gorilla's mind.

In effect, my career in psychology has been one of climbing the primate ladder—if in fact we can consider one primate higher than another. I began at Stanford working on a study of attachment behavior in rhesus monkeys under the guidance of Karl Pribram, a leading theorist on neuropsychology. In this study, infants were separated from their mothers (briefly) to prove what seemed to me the obvious point that they would prefer their mother to a peer and a peer to an empty cage as a source of comfort in an anxiety-producing situation.

Next I became involved in a study of self-recognition in gibbons. Simply put, this means that I was trying to see whether a gibbon knew whom it was looking at when it saw its image in a mirror. I found this study more intriguing because it would indicate whether the ape had any consciousness, a quality that until recently was thought to be uniquely human. A 1970 study had proven chimps capable of self-recognition, and the purpose of the study of gibbons was to help to determine how far down the evolutionary scale this ability might extend. The six months of the study produced no signs of self-recognition in the gibbons.

It was shortly after I began work on the self-recognition study that the Gardners came to Stanford to speak. From that moment onward, I began looking for opportunities to work with a chimp, or failing that, any great ape. Thus I agreed instantly in September 1971 when Karl Pribram suggested that I accompany him to San Francisco to look at the gorilla colony there. Dr. Pribram was toying with the idea of constructing a sturdy console with an encoded keyboard connected to a computer, which he would then use to teach the gorillas to communicate by pressing different keys.

When we arrived at the San Francisco Zoo, we met the director, Ronald Reuther, and then walked over to the gorilla grotto, a large, rocky, cement area separated from onlookers by a dry moat. While Dr. Pribram and Mr. Reuther discussed the pros and cons of the proposed experiment, I became absorbed watching the gorillas idly pass the day. The tableau was a study in lassitude, broken only by a little struggle between a mother gorilla and her infant. The tiny gorilla was clinging ferociously to its mother, who kept pushing the baby up onto her back, only to have the baby slide off each time. The sight of the infant brought my mind back to my quest. I was not that interested in Pribram's proposed experiments because I had already concluded from my reading on the subject that a sign language was the most productive way to study ape language abilities. As I watched the infant I thought, "Well, Pribram can have his experiment, and I will just have mine with this baby." It did not turn out to be so simple.

When I made this proposal to the zoo director, I was turned

down. A primary goal of the zoo was to breed endangered species such as the gorilla, and Mr. Reuther, sensibly enough, felt it would not advance that purpose to separate the infant from its mother at the tender age of three months. Undaunted, I continued my study of Ameslan and resolved to find another gorilla or wait until this infant was older. I tried to find out what I could about the baby gorilla and her circumstances at the zoo.

The infant was Koko. The mother who had so peremptorily placed her daughter on her back was Jacqueline, nicknamed Jackie. Poor Jackie had previously suffered the indignity of being thought to be a male. In fact, she had been purchased from the Brookfield Zoo in Chicago to be the mate for Missus, one of the San Francisco Zoo's female gorillas. Jackie came to San Francisco courtesy of Carroll Soo Hoo, a philanthropic businessman, who donated the money to purchase Jackie—then named Jacob—and another gorilla. The zoo expectantly closeted Jacob with Missus and nervously wondered why the couple did not hit it off and raise a family.

Ultimately, the zoo discovered their error and, with some embarrassment, decided that the cause of breeding gorillas might be better served if Jacob was put in with a male. The zoo changed her name to Jacqueline, and undoubtedly to her vast relief, Jackie was introduced to Bwana. They did mate, and a female gorilla was born on the Fourth of July in 1971. The zoo held a contest to choose a name for the infant. The winning entry was Hanabi-Ko, Japanese for "Fireworks Child."

The zoo's plan to keep Koko with her mother did not work out as they had hoped. Shortly after Dr. Pribram and I visited the gorilla grotto, Koko's health began to deteriorate. Jackie was a good mother, but the San Francisco Zoo is not the jungle. Jackie's milk was not sufficient to keep Koko nourished, nor could Koko supplement her mother's milk with forage as infant gorillas are reported to do in the wild. She became undernourished, and when an outbreak of shigella enteritis swept through the gorilla compound, she almost died. Suffering from malnutrition, racked with diarrhea and septicemia, hairless, and dehydrated, Koko was a pathetic 4 pounds 14 ounces—the average birth weight of gorillas—at the age of six months. At that point, just before Christmas, Koko was separated from her mother and taken to the Animal Care Facility of the University of California Medical Center in San Francisco for a few days before being taken into the Reuther household for two weeks. With round-the-clock care, she recovered sufficiently to be transferred to the house of Deedee and Landis Bell, manager of the Children's Zoo, on the Children's Zoo grounds. After six months in the Bells' care, the zoo felt it was time to put Koko back on permanent display, and installed her in the nursery of the Children's Zoo. Subsequent examinations determined that she had suffered no discernible lasting harm as a result of her illness.

At about this time, I made another trip to the zoo. I had come up to photograph gibbons as part of the self-recognition study. I ran into one of the keepers, Marty Diaz, who told me about Koko's illness. He suggested that the zoo might now listen more favorably to a proposal, if I still wanted to work with Koko.

Marty Diaz was most sympathetic to my desire to work with sign language, and he offered to speak to Mr. Reuther on my behalf. That same day, I asked my advisor for permission to switch to a language project with Koko. Mr. Reuther and my advisor both granted their permission, and the very next day, with no funding, few private resources, and as yet no formal project design, Project Koko began.

Too excited to be tired from a night sleepless with anticipation, I drove from Stanford to San Francisco with Ron Cohn to meet Koko on a foggy Wednesday morning on July 12, 1972. When I entered the nursery of the Children's Zoo, Koko left the arms of her caretaker, Debbie Lee, for mine. She pushed her soft face close to mine, smelling me and looking me over. Then Debbie put the 20-pound gorilla, all black save for a white rump patch, onto the nursery floor and I signed, *Hello* (a gesture somewhat like a salute). Koko put her hand on her head and patted it and then promptly pulled my hair as I sat down.

The glimpse I had caught of her sleeping serenely in her basket the day before did not prepare me for this interaction— she was a real dynamo and seemed much bigger this day. While Debbie was in the room with us, Koko responded to my beckoning *come* gesture, but later, when alone with me, she went on about her play with her toys as if I wasn't there. Whenever I stood up, however, she rushed to my feet and started to scale my legs—evidently she thought I was leaving. At one point Koko became excited and played a game of peekaboo behind a door with Ron when he and Debbie joined me in the room for a quick photo session. Later, while Debbie and I chatted, Koko bit me a couple of times. Taking this as a sign that we had perhaps overstayed our welcome, Ron and I departed for the day.

The next morning I arrived at 9:00 with a wading pool for Koko. She cautiously put her nose up to it, touched it, and nibbled on the edge. When Debbie placed Koko in it she immediately ran her fingers over the upraised bubbles on the bottom of the pool. She delighted in running in and out of it and splashing in a few inches of water. Excited by the pool, she nipped me several times, but by now I was learning to anticipate and divert these testy assaults.

While the zoo volunteers performed the morning chores I joined Koko in the nursery. She still ignored me often, but when the horses, goats, and sheep were let out into the zoo yard and stampeded by the nursery window, Koko scrambled over to me and briefly clung to my clothes. Then the whirring of the blender to mix her formula of similac and strained cereal set her into a frenzy of activity: She vigorously banged her toys around, and repeatedly pounded on and rolled herself over a rubber dog. She interrupted her wild play only to peek under the door to the adjoining room where her bottle was being prepared and to hammer on the door periodically. I asked the zoo volunteers to sign *drink* before feeding Koko her formula and *up* before picking her up.

Initially, Koko seemed to prefer men to women. During the first week, she was more inclined to interact with Ron and my office mate, John Bonvillian, than with me. She took to John very well—examining his beard closely, sniffing, fingering, tasting, and yanking it. She climbed all over him jungle-gym style and rode on his back. Ron also got the jungle-gym treatment, and Koko was very responsive to him. She imitated his twisting of a knob on her toy clock, and his clapping. When my male friends were present, Koko interacted very little with me. I also somewhat enviously noted that she never attempted to bite them. After a couple of weeks, though, she seemed to conclude that I was reliable, and likely to be a permanent fixture in her life. She attempted to bite me less and less frequently, and she also began to show a preference to be held by me rather than by a man when she had an alternative. Her first response when frightened was to jump into my arms and cling tenaciously.

From the beginning of Project Koko I had a dual role: I was a scientist attempting to teach a gorilla a human sign language, but I was also a mother to a one-year-old infant with all an infant's needs and fears. My initial problem was to establish rapport with Koko, who was, perhaps because of the unsettling events that had marked her short life, at first suspicious of this strange blonde human.

Each morning before the zoo opened to the public I would carry Koko for walks through the Children's Zoo. I felt it was important to get Koko out of the confines of the nursery at every opportunity. At first I had no need to restrain her with a leash; for one thing, it is normal for an infant gorilla to stay on or near its mother for the first year and a half of life, and for another, Koko was terrified of the large animals (particularly a baby elephant who was fond of trumpeting every morning) and wouldn't venture from my side. The only large animals that Koko could intimidate at that age were a herd of surpassingly stupid llamas. They would congregate at the fence when we passed, apparently under the impression that we were zoogoers bearing llama food. Koko would rush at them threateningly and enjoy with evident satisfaction the stampede she precipitated.

One animal Koko was particularly afraid of was the gorilla. When I took Koko on a trip to see her parents at close quarters inside the gorilla compound, her relatives gathered quietly to examine the little gorilla. Bwana, the dominant male and protector of the group, was upset when he first saw us approach; he barked, followed us, and threw feces at us. Frightened, Koko squirmed and defecated in my arms. We left in a hurry.

With the beginnings of our rapport, the problem was to focus Koko's attention on hands. Koko was, after all, only one year old, and when not asleep, she was constantly moving and exploring. I would construct little games to divert her and show her the utility of her hands. I breathed fog onto the glass of the large windows in her room and then drew stars and simple faces on the misted surface. Koko loved these games and would attempt to draw as well, although what appeared were amorphous squiggles.

It was impossible within the confines of Koko's display cage to seal her off from spoken language (the glass was hardly soundproof and some zoo visitors seemed to take it as a sacred obligation to make remarks to Koko and whoever was in with her). Consequently, I decided to make a virtue of necessity by adopting a method known as "simultaneous or total communication." This simply means that the speaker accompanies his signing with the spoken equivalent of the message.

The ambitions of the project were quite modest at first. On July 22, Karl Pribram and I spoke with Ronald Reuther about the amount of time I was to be allowed for the project. Mr. Reuther's idea was to reunite Koko with the other gorillas as soon as possible, which he thought would be in about six months. On the other hand, Landis Bell, the director of the Children's Zoo, thought Koko should not be put back with other gorillas for about three years. I was a bit disappointed at this point, since I hoped to carry on my work with Koko for as long as the Gardners had worked with Washoe—four years. On the other hand, Dr. Pribram felt that I should concentrate on teaching Koko only three or four signs. I thought she could probably handle more than that, but decided to begin by molding and shaping *drink, food,* and *more.*

I would divide Koko's bottle into two portions, and would sign *drink* before giving her each portion. The *drink* sign is made by shaping one's hand somewhat like a hitchhiking gesture, and then placing the extended thumb to the lips. While preparing and offering the bottle, I made this gesture, and then attempted to get Koko to make the gesture. Koko, being a one-year-old, had few thoughts other than getting her hands on the bottle, and then the bottle into her mouth.

Although I tried for a strict routine, we were frequently interrupted when children came up to the window and dangled toys or food on the other side of the glass. Koko would rush up to try to touch or mouth the objects through the glass, and then, when she discovered they were out of reach, she would pound on the glass in frustration. Her principal amusement those first few weeks was to close her eyes and spin wildly around the cage—something gorillas do in the wild. As Koko grew older, she embellished this game by pulling a blanket over her eyes, generally when she had some mischievous intent, such as giving a playful smack to a human companion. Possibly Koko felt that by pulling the blanket over her eyes she became invisible. Indeed, she was perpetually surprised to find herself accused of these petty assaults.

During the first few months of the project, the Children's Zoo volunteers who had looked after Koko before my project began continued to look after her when I could not be with her. At the end of the first summer, these volunteers had to go back to school during the week, but I was able to fill the gaps with two new volunteers who offered their services. One was a deaf woman, the mother of my sign language teacher. The other was Barbara Hiller, a docent at the zoo. Barbara cared for Koko from the time she was in diapers and is still with the project today. Later in the fall, the Stanford psychology department provided salary money that permitted me to hire Hank Berman, an assistant whose native language was sign.

As Project Koko got underway, I had the advantage of surveying the trial-and-error approach to teaching language used in previous experiments with chimps. These experiments also produced a great fund of information against which I might judge Koko's performance—if, in fact, she learned language at all. In 1972, when I began Project Koko, there were a great number of scientists who disputed that the chimps' achievements had any linguistic significance. Project Koko began during turbulent times in the behavioral sciences, and it was only because of previous pioneering work with chimps that I had any chance of being taken seriously. My cause was not helped by the fact that the subject of my experiment was not a chimp, but a gorilla.

3

Gorilla Gorilla

If some of my colleagues were skeptical of my ambitions to teach a gorilla sign language, it was partly because of the gorilla's reputation for being ferocious, stubborn, and stupid. While chimps have traditionally been the teacher's pet of the behavioral sciences, the rare, self-absorbed gorilla has been given a wide berth by scientists mindful of the animal's strength. Throughout the century timorous researchers have justified this neglect by reciting like a catechism a literature on the animal's intractable nature and dubious intelligence.

The gorilla, as every reader knows, has not had a good press. Part of its problem is that the gorilla does not have much documented history. Creditable sightings only date from the mid-nineteenth century. Early accounts spoke of the animal's ferocity and enormous strength. One hunter reported that an enraged gorilla grabbed his gun and crushed the barrel with his teeth. The French-American explorer Paul du Chaillu probably did most to create the popular image of gorillas that still persists today. Du Chaillu caught the public imagination with his lurid description of a gorilla kill in 1861: "His eyes began to flash fiercer fire as we stood motionless on the defensive, and the crest of short hair which stands on his forehead began to twitch rapidly up and down, while his powerful fangs were shown as he again sent forth a thunderous roar. And now he truly reminded me of nothing but some hellish dream creature—a being of that hideous order, half-man half-beast, which we find pictured by old artists in some representations of the infernal regions...." The legacy of such reports shows in a recent poll of British schoolchildren: gorillas ranked with rats and spiders as the most hated and feared creatures on earth.

Given the gorilla's awesome image, many people asked me how I would dare to enter the cage of an animal that so terrorized the brutish hunters of the last century. For one thing, I had read another body of scientific literature that described an entirely different animal from the hellish creature of the popular accounts (although even some scientific writings fell prey to superficial prejudices based on the gorilla's appearance). According to George Schaller and Dian Fossey, who have studied gorillas in the wild, they are peace-loving vegetarians despite the intimidating chest-beating, branch-breaking, and roaring displays they may use to greet intruders. They roam the forests of Central Africa in nomadic bands of some two to thirty individuals led by a dominant older male. Their communication consists of a combination of postures, gestures, and vocalizations. A sideways glance and an annoyance bark from the dominant male are usually enough to resolve disputes; a grunt or purring vocalization indicates contentment and social harmony. The life span of gorillas in the wild is not known; conservative estimates place it at thirty. In captivity gorillas have been known to reach fifty. In physical development, a ten-year-old gorilla is roughly equivalent to a twelve-to-fifteen-year-old human. Although females are willing to mate from about age seven to nine in the wild (about six in captivity), they usually do not conceive until age ten or eleven (seven to ten in captivity). Males are ready to mate at about age nine or ten. The female often initiates courtship when she is in estrus, and the male usually indicates interest only then. Gorillas in the wild tend to spend much of their time lolling about, eating several times a day from a ready supply of vegetation; and, except for man, they have no enemies.

These firsthand reports of the gorilla's gentle nature, along with the photographs Carroll Soo Hoo had often shown me of himself roughhousing with Bwana and other 200-pound gorillas, were enough to still any doubts I might have entertained about the dangers of working with Koko.

Contrary to its popular image, the gorilla is less aggressive, less excitable, and in some ways a good deal easier to work with than I had anticipated. That this is not better known is partly because the gorilla is very difficult to obtain for research. But I suspect that many researchers would rather not risk giving a 400-pound animal the benefit of the doubt that is necessary to find out what the animal is really like. Most, if given a choice, would probably prefer to work with chimps, who genuinely seem to enjoy the company of humans. Roger Fouts, a psychologist who has extensively studied chimp use of sign language, remarked that he did not like the way gorillas hunker down at a forty-five-degree angle, turn their heads, and stare sideways at him. Because so little work has been done with gorillas, they have been unfairly regarded as an intellectually disadvantaged, moody, and uncooperative poor relation of the great apes.

Gorillas are great apes, a term that refers to the family Pongidae, or pongids. It includes the orangutan (*Pongo*), the chimpanzee (*Pan*), and the gorilla (*Gorilla gorilla*). The orang, or red ape, is a native of Borneo and Sumatra, while the chimp and gorilla are now found only in an ever-diminishing band that runs through equatorial Africa. All three are threatened in the wild by habitat destruction, hunting, and what is euphemistically called "collection" for zoos and laboratories. There are only some 250 mountain gorillas left in existence; lowland gorillas number fewer than 5,000 or 10,000. It is unknown whether the gorilla was ever particularly abundant, but its existence, in spite of recent laws to protect it, is now possibly the most precarious of all the great apes'.

Together with the lesser apes (the gibbon and the siamang) and man, the great apes are members of the superfamily Hominoidea. Hominoidea, in turn, is a part of the suborder Simiae of the order Primates.

Scientists have long debated over which of the great apes is man's closest relative. Depending on whom you talk to and what aspect of the ape's physiology is being examined, researchers make varying claims for the chimp or the gorilla.

Adding to this confusion is the assertion by some scientists that the orangutan's brain most closely resembles man's in certain anatomical ways related to the evolution of language. This is somewhat surprising, because the orang is commonly regarded as man's most distant relative among the great apes. For the moment, the question of which ape is most closely related to man will have to be considered open because of the lack of comparative data.

Also unsettled is the issue of which great ape is the most intelligent. Such a question is somewhat charged, since we would hardly be comfortable if our closest relative turned out to be something of a dolt compared with the other two. For a long time it was generally assumed that the chimp was the brightest, although there is little hard data to back this up. In fact, as primate psychologist Duane Rumbaugh has pointed out, when people are asked how they know the chimp is bright, many will cite the descriptive tag on the chimp cage at the zoo. Because we consider the chimp our closest relative, we have tended to accept its intellectual superiority over the gorilla without too much scrutiny. And since chimps are the easiest of the three great apes to test for intelligence, the claim tends to become a self-fulfilling prophecy.

Before Project Koko got underway, Duane Rumbaugh administered a series of tests to determine the relative intelligence of a group of chimps, orangs, gorillas, and a pygmy chimp. The tests were inconclusive. One orang consistently had the highest scores. But Rumbaugh wondered how significant the gorilla's low scores were, since it frequently disrupted the test and ultimately crashed the test apparatus.

Later learning tests were more conclusive. Required to discriminate between different objects according to varying criteria, the gorilla and orang both performed better than the chimp. Rumbaugh believed that his data had at least exploded the myth of chimp intellectual superiority among the great apes. He remarked that the difference in mechanical aptitudes, or simply in how the animals felt on a particular day, may have had a lot to do with the differences in their performance. (To this observation I say amen.) Rumbaugh noted that vocabulary size would probably be the most reliable measure of intelligence, but since he conducted these tests before it was believed that the apes might develop any vocabulary at all, he had to conclude ruefully that the question was, for the time being, moot. The breakthrough in communicating with chimps, orangs, and gorillas has fostered a renaissance in the study of ape intelligence. I will come back to the issue of intelligence later.

It was probably because of behavior like that of Rumbaugh's gorilla destroying the test apparatus that the gorilla developed its reputation as difficult. Two researchers, Hilda Knobloch and Benjamin Pasamanick, went so far as to claim that the gorilla was uncooperative because it was stupid: "There is little question the chimpanzee is capable of conceptualization and abstraction that is beyond the abilities of the gorilla. It is precisely because of these limitations, which are apparently genetically determined...that it is more difficult to work with them." The great primatologist Robert Yerkes shared some of these feelings, but he also suspected that the gorilla's intransigence might indicate the presence of intelligence rather than its absence. In 1925 he wrote, "In degree of docility and good nature the gorilla is so far inferior to the chimpanzee that it is not likely to usurp the latter's place...in scientific laboratories." It also occurred to Yerkes that the gorilla was "a natural experiment in which the value of brawn versus brain is being determined." Ultimately, however, Yerkes's clearheaded understanding of his beloved apes led him to observe, "It is entirely possible that the gorilla, while being distinctly inferior to the chimpanzee in ability to use and fashion implements and operate mechanisms, is superior to it in other modes of behavioral adaptation and may indeed possess a higher order of intelligence than any other existing anthropoid ape."

Today, more than fifty years after Yerkes made his remark, Koko's performance bolsters his intuition.

4

Tumultuous Times

The attempts to communicate with apes have been marked by controversy from the time of the first successful attempt to converse with another animal. The problems are twofold. First of all, the idea that language is what separates man from animal is enormously important to the way we view and act in the world, and is not the type of concept that can be cast aside blithely. Secondly, it is one thing to seem to converse with another animal, but it is quite another to be able to prove that the animal's responses are not simple mimicry or trickery. After all, stories about "talking cats" or "talking dogs" inevitably turn out to be whimsy. Why should anyone take the notion of "talking apes" any more seriously? The difference is that the work with apes has involved experiments designed in such a way as to isolate different aspects of language and to rule out alternative explanations of what the ape is doing when it uses the language. Such rigor was necessary at the beginning of these experiments if the idea of conversing with an ape was to be perceived as anything other than wishful thinking, and that rigor has been necessary throughout Project Koko.

In July 1972, when I began to work with Koko, there was already a body of literature that suggested vastly greater capacities for language in the great apes than the few meager spoken words several previous experiments had produced. This was chiefly due to the Gardners' work with Washoe.

It was the Gardners' insight to design an experiment that

21

separated the concept of language from speech. The subject of the Gardners' experiment was Washoe, a wild-born female chimpanzee whom they began training in June 1966. The Gardners began their work at a time when scientists were citing an elaborate attempt by Keith and Catherine Hayes to teach spoken language to a chimp named Viki as conclusive evidence that language was the critical ability that separated man from the other animals. Viki had proved almost the peer of normal human children in performing a number of perceptual and analytical tasks, but she was never able to speak more than five or six words, and she uttered these simple words only with great difficulty. When the Gardners saw films of Viki, they watched with a different eye from those who assumed that Viki's limitations were due to mental inadequacy. They noted that the chimp was almost intelligible without the sound track, and that she consistently made characteristic gestures when she tried to speak. Moreover, Viki learned to say only words like "cup," which she approximated by reproducing the unvoiced "c" and "p"; she was never able to voice the "u." The Gardners began to wonder whether Viki's problem was physical rather than mental. They decided it would be worthwhile to test a suggestion made by Robert Yerkes fifty years earlier—that sign language might be the most productive medium for establishing communication with the apes.

Because they were pioneering, the Gardners had no precedents to guide them. They were not sure which teaching method might be the most productive with a chimp, and they were not sure what language to use—an existing sign language or a gestural language they might invent. They chose to teach Washoe Ameslan because it was well known and had been studied, and because it would allow them to compare Washoe's performance with that of deaf children and normal speaking children.

After experimenting with various methods deriving from different theories of language acquisition, they settled on an instructional method called "molding," in which the teacher takes the subject's hands and forms them into the proper configuration for a sign while the child or chimp looks at some representation of what it signifies. The Gardners were not doctrinaire about this method, however; if Washoe picked up a sign through imitation, or through the progressive "shaping" of her gestures, the Gardners would exploit these opportunities as well.

By the end of only twenty-two months of training, Washoe had acquired 30 signs that she used "spontaneously and appropriately." Her vocabulary was four times larger than the largest acquired by any other ape in the experiments using spoken language.

Because the Gardners were conducting these experiments at a time when the behavioral sciences were generally hostile to the idea that an animal might learn language, they had to be above suspicion in their methods of data collection and testing. To prevent the possibility of cueing Washoe (inadvertently giving her the answers), they used a method of double-blind testing, in which the tester could see what Washoe was signing but could not see the object that elicited the sign.

Perhaps the clearest evidence that Washoe was something more than a clever mimic was the way she seized on the utility of the language. She invented a sign for *bib* which the Gardners rejected but which, upon examination, turned out to be the correct gesture in Ameslan. She also invented a sign for *hide* that she used to initiate one of her favorite games, hide-and-seek.

It was not only Washoe who was shaking up ideas in the scientific establishment. At about the same time that the Gardners began to publish their findings with Washoe, David Premack, a behavioral psychologist from the University of California at Santa Barbara, began to publish the results of his attempt to teach a female chimp named Sarah an invented "language" using plastic tokens placed on a magnetized board.

Given the modesty of the Gardners' published claims for their subject, the response was extraordinary. Their news that one animal had used a human language precipitated a thunderstorm of criticism from many eminent scientists who had already gone on record saying that animals could not learn language. The Gardners had merely presented a list of two-word phrases generated by Washoe and claimed that Washoe's early sentences compared with the early sentences of children of equivalent age. They noted that they fully expected the child to outpace the chimp in language acquisition eventually, and said that they simply wanted to determine at what point this occurred. They were not trying to show that Washoe had mastered language, but only that there was continuity between animal and human communication.

If the Gardners were trying to show that in some ways Washoe communicated like a child, they were criticized as though they had said that the chimp was the next Mark Twain. Immediately after the first publication of their findings in 1969 in *Science*, rebuttals began to appear, written by the most distinguished names in the behavioral sciences. Roger Brown, one of the first psycholinguists; Erich Lenneberg, another distinguished psycholinguist; geneticist Theodosius Dobzhansky; and Jacob Bronowski, among others—all disputed that Washoe had acquired language. Some, like Dobzhansky, went out of their field to attack the Gardners' findings. It is unclear whether Bronowski left his field, because he had so many.

Bronowski and Ursula Bellugi, a distinguished psycholinguist who was then a graduate student, wrote a brilliant exposition about the structure of the sentence and then offered a laundry-list of reasons why Washoe did not have language. This list— Washoe did not ask questions, she did not say no, she had no sense of word order—like so many of the criticisms leveled by others, turned out to be premature. Bellugi subsequently took back her criticism—indeed, many of the early conclusions were later recanted. It is ironic that long after Bellugi revised her early criticisms of Washoe, other scientists were still citing her original article in support of their skepticism of the sign language experiment.

The Gardners' response to these objections was merely to ask: How can one be so sure that Washoe does not have language when there is no agreement among linguists about what language is or when a child can be said to have it?

The fact that so many eminent scientists hastily dismissed the language experiments with chimps was not simply because they had earlier written that only man had language, and that they hated to admit they were wrong. Rather, their reactions illustrated a basic truth about the nature of scientific change: Science can discover that something is wrong with its guiding principles (for instance, the ancient idea that the earth is the center of the universe) only if scientists are passionately and rigorously dedicated to those erroneous principles. Using those principles, the scientist will pursue investigations into the unknowns of his science. If something is wrong with those guiding principles, his research will at some point turn up anomalies (eccentricities in the orbits of the planets, for instance) that either cannot be explained by the principles of the science or might be more economically explained by another set of principles (by installing the sun as the center of the solar system). When the alternative explanation of the anomaly appears, science does not then change by mass conversion. Rather, adherents of the old idea and the new idea exist side by side for a time, until ultimately those holding the old idea die out and are succeeded by scientists educated under the new view of things. Thus science proceeds by revolution. This in a nutshell is the model for scientific change proposed by Thomas Kuhn.

Although Kuhn based his model on the so-called hard sciences such as physics and biology, it would seem to explain the somewhat confusing situation that surrounds the language experiments with animals. The difference is that in the behavioral sciences the lines between philosophical and scientific principles are much more blurred. Not only has the idea that man is the only animal capable of language been argued by scientists, but it also appears in the Bible, in interpretations of the Bible, and throughout Western history in different philosophical tracts. The notion that only man has language is bound up with arguments involving our rights to experiment with or harvest natural resources, and indeed forms the basis for the development of Western civilization. Therefore, it is hard to find an aspect of life in a modern society that does not at some level touch on the question of whether or not language separates animal from man. Since the argument for human uniqueness that was threatened by the anomaly of the Gardners' success with Washoe is one of the most pervasive tenets of modern life, it should not then be surprising that there was a large constituency of eminent scientists who were committed to the notion that animals cannot learn language, and were willing to revise their definition of language to keep upstarts like Washoe and Koko out of language's exclusive club. Nor should it be surprising that this debate continues fifteen years after the first animal conversed with a human in a human language.

The curious thing about the devotion to the anti-evolutionary notion of man's language uniqueness is that some of the great evolutionary scholars of our times sedulously adhere to it. In a world in which we see graceful continuities linking us anatomically and behaviorally to the rest of the animal kingdom, language theories require us to accept an awkward discontinuity when we consider communication. All the Gardners were asking was why, if there is continuity in every other aspect of anatomy and behavior, should there not be continuity in communication. The answer turns out to have to do with a lot of things other than language.

This was the turbulent climate of the behavioral sciences in which Project Koko began. I could profit from the methods and experience of the Gardners and, because of them, did not have to refight the initial battles for credibility. However, a significant number of behavioral scientists still considered these interests odd, if not heretical. And when some Stanford psychology professors noted that I was enjoying myself, their attitude was "When are you going to stop fooling around with gorillas and start doing some serious work for your thesis?"

5

Koko's First Words

Koko first began to show signs she understood the significance of the strange gestures she was constantly witnessing as early as the second week of Project Koko. On July 25, before Koko had been taught any signs through molding, the volunteers reported that she made gestures that resembled the *food* and *drink* signs several times during the morning before I arrived. I was reluctant to accept this as significant. Koko was not making the signs spontaneously in my presence, and I had no reason yet to accept that she was learning by observation alone. (In retrospect, I believe that Koko probably did try to make these signs; she has subsequently surprised me often by making signs she has learned only by observation, without any active instruction.) The volunteers continued to report what seemed to be signing attempts, and I began to notice that Koko was starting to use "natural" signs observed in wild gorillas, such as *gimme*, which looks like a beckoning gesture.

Over the next two weeks, Koko continued her spontaneous approximations of signs, but to me they seemed coincidental, random, unintentional. With all her fidgeting, I wondered whether any of our intent was getting through. On August 7, we began a formal routine of active instruction. My assistants and I used every opportunity that arose during the day to teach Koko *food*, *drink*, and *more*. Rather than hand her her bottle as a matter of course, we would first hold it up and let her see it. If she responded by signing *drink*, we'd give her the bottle. If she made no response, we'd sign, *What's this?* If that still elicited no response, we'd mold her hand into the sign for *drink*. I also asked the zoo volunteers to include some signing in their daily caretaking routine when my assistants and I were absent.

Only two days into this training routine Koko said her first word. On August 9, she consistently responded with close approximations of the *food* sign when I offered her tidbits of fruit. Most frequently she put her index finger to her mouth, but she also made the sign correctly—putting all the fingers of one hand, held palm down, to her mouth. As it dawned on me that for the first time she was consistent and deliberate in her signing, I wanted to jump for joy. Finally she seemed to have made the connection between the gesture and the delivery of food, to have discovered that she could direct my behavior with her own.

I praised Koko profusely and seized every chance to get her to sign *food*, showering her with treats in the process. Whenever she reached for some food, I would prompt her by signing *food*, and almost every time she responded. I made sure that she realized she was supposed to ask for things by name by pushing her hand away and signing *no* when she did not make the sign. On several occasions Koko signed *food* without any prompting on my part. After her nap I gave Koko another twenty or so opportunities to sign *food*, and she responded incorrectly only toward the end of the afternoon, by which time the stuffed gorilla had no interest in food whatsoever.

I could not wait to share the news of Koko's breakthrough with Ron and my friends in the Psychology Department at Stanford. Koko too seemed to realize that something exciting was occurring. She was agitated all day, and at one point during the afternoon, she put a bucket over her head and ran around wildly.

Although Koko did not immediately go on to ask the names of other objects, she did attempt to extend the use of her new sign to other situations the next day. She repeatedly used the sign as she watched a volunteer removing discarded food while cleaning her (Koko's) room.

Once Koko made the association between her hand gestures and the objects they represented, she quickly learned the words *drink, more, out, dog, come-gimme, up, toothbrush,* and *that.* Barely into the second month of training, she moved from one-word expressions to two-word combinations—somewhat more quickly than Washoe had. Washoe's first reported combination occurred in her tenth month of training, when she signed *Gimme sweet*. Koko, on the other hand, signed *Gimme food* on August 14, 1972, but because the *gimme* sign in this case might have been a natural reaching motion that Koko combined with the sign for food, I couldn't accept her gesture as a legitimate two-word combination. However, before any doubts about Koko's precocity in combining could arise, she followed up by signing *Food drink* eleven days later. She used this to describe her formula, a mixture of cereal and milk. About a month later, Koko said, *Food more*, to ask for more fruit during a teaching session.

In all, during the first two months, Koko used about 16 different combinations, most of which were limited by her small vocabulary to requests for food or drink: *More food, Drink there, More drink more, There mouth, mouth-you there*, and *Drink more food more*. I accepted about one-third as legitimate expressions of semantic relations.

One of the early criticisms of Washoe, later refuted, was that she did not ask questions. By the third month—September— Koko began to ask questions as well, although she would not phrase them the same way chimps did. Washoe, Lucy, and other sign-language-using chimps were taught to make the sign *question*, which is to simply draw a question mark with the forefinger in the air in front of one's body. Instead of this, from the outset Koko spontaneously used eye contact and gestural intonation to phrase questions, a form that is considered legitimate in Ameslan.

I first noticed it on one afternoon in early September. I was blowing on the window and urging Koko to draw in the mist. After I demonstrated, she did. Then she pointed to my mouth and touched it with her index finger while looking into my eyes. I assumed she was asking me to blow again, and I enthusiastically complied. Soon she tried making her own fog by putting her mouth close to the window, opening it, and extending her tongue slightly, almost licking the window. She succeeded in creating a bit of a mist and drew in it with her finger. Later that day she even more closely approximated my fog-making by adding the hah-hah sound I made when blowing on the window.

A week later Koko made a more elaborate request. As a couple with an infant approached the window, Koko pointed to the glass, then to her mouth, then to my mouth, and then to the glass again. She immediately repeated this same sequence and looked into my eyes. Surprised and fascinated by the complexity of her request, I took a few seconds to guess that she wanted to play the fog-blowing game. I huffed a mist and she drew in it. Then Koko again tried to make her own fog by putting her mouth and tongue to the window.

In addition to making requests, Koko began to give an interrogative cast to signed phrases. By cocking her head, raising her eyebrows, and maintaining eye contact, she turned *There food* into a question as she was being carried off to the nursery, and used the same expression to ask *You there?* while pointing to the glass window.

As Koko's language skills developed, so did her physical coordination and mental sophistication. In October, when she was fifteen months old, her motor skills were rapidly improving and her perceptual abilities becoming very sharp. She figured out how to turn on the kitchen faucet to get herself a drink, made serious but uncoordinated attempts to return the spoon to a container of yogurt in order to feed herself, and manipulated four wooden sticks simultaneously in play.

As much as she enjoyed our dexterity exercises, however, she could not be tricked, even by Ron's clever schemes, into contacting objects she feared. Once Ron attached a rubber spider Koko hated to a large plastic bead with a clear fish line. While I worked on the signs *dog* and *baby* with Koko, Ron placed the bead under the door to Koko's room, hiding the spider out of sight around the corner. Koko saw none of the preparations. When she noticed the bead, she went over to it but looked under the door before pulling on it. The spider came into view, and she jumped back. Ron hid the spider again and Koko pulled on the bead twice more, recoiling both times the spider emerged. After this she batted the bead away when it was presented.

This same day Ron distracted Koko during a feeding session by curling his tongue. She watched him intently through the screen mesh partition in her room and started moving her tongue around in her closed mouth. When Ron left, Koko pounded on the screen until he returned to repeat the performance. Later Koko did something simple but somehow very touching. She took me gently by the hand and led me around her room, pausing frequently to adjust the position of our hands.

If Koko's dexterity was improving, there were still significant limitations on her physical capabilities for signing. A gorilla's hands are somewhat different from the average child's. They are bigger, of course, but they are less well organized for precise motor tasks than ours. The thumb is smaller and placed farther down the hand and away from the rest of the fingers than a human's is. Moreover, the gorilla's precise motor control over its hands, while considerable, is less well developed than ours. This means that certain signs are difficult for Koko to form. In these cases either she will adapt the sign herself, or we will invent a variant for her. For instance, *water* is made by touching the finger-spelling of the letter "W" to the signer's lips. Since Koko cannot make a "W" with her hands (her thumb won't reach her little finger), she will touch the side of an extended index finger to her chin. Similarly, *sand* and *purple* are physically impossible for Koko to articulate because of the small size of her thumb.

Until age four, Koko had trouble executing signs made away from the body, which was true of Washoe as well. Perhaps it was because signs made by bringing hands into contact with the body are better grounded or oriented than those made in the air. Both Koko and Washoe acquired touch signs more rapidly than non-touch signs, although there is no conceptual difference in signs made away from the body. Koko even tried to convert non-touch signs into touch signs by making them on the body rather than in front of it. *Finished*, for instance, is made with both hands out in front of the body, about shoulder width apart. The hands are held vertically, thumbs up and palms facing the body, and shaken. Koko used to make the *finished* sign by shaking her hands against her chest. Similarly, the sign *milk* involves holding one fist out in front of the body and then squeezing, as if milking a cow. Koko knocked her chest with her fist to say *milk*. (Now, however, she articulates both signs properly.)

Another curiosity of Koko's signing, probably also related to her preference for signs that make contact with the body, is her habit of making motion signs (such as *long*) starting close to the trunk of the body and moving away rather than the other way around. This reversal has been noticed in autistic children as well.

Not all of Koko's variations, mistakes, and inabilities stemmed from physical limitations. In trying to sort out physical from intellectual influences on her signing ability, I saw that she often made common "baby" errors. Deaf infants use a form of baby talk which may invert the motion or simplify the form of a sign. When babies are learning a sign, they have to generate the mirror image of what they are looking at. Many babies do not complete this adjustment for some time. The sign *bird* is made by forming the index finger and thumb into a configuration somewhat like a bird's beak and then placing the hand beside the mouth pointing outward. Koko makes this gesture with the fingers pointing toward the mouth.

Another important influence on Koko's growing signing ability was simply her motivation. Both Washoe and Koko quickly learned signs for objects or actions they desired. Washoe picked up *lollipop* without direct instruction, and Koko similarly learned *swing* and *berry* by imitation within minutes. On the other hand, she took months to pick up the sign for egg, a food she dislikes.

Koko was often sloppy in her signing and would elide one sign into another, or reduce a gesture to its barest skeleton, but in this she was not unlike fluent signers in Ameslan. When two fluent signers are talking, they may frequently take some of the same shortcuts that Koko did. Anyone will recognize that this is the case with spoken language as well. Few people clearly enunciate grammatically precise English. In fact it sounds strange when you hear it. Rather, what are called paralinguistic phenomena—such as cadence, intonation, gesticulation, and stock abbreviations—bear a large measure of the communicative burden. A conversation between two people who know each other well can sound like a meaningless series of mumbles and monosyllables.

Koko's vocabulary was growing at about the same pace as Washoe's—one new sign learned each month—for the first year and a half. At the end of eighteen months, Koko had acquired 22 signs, about the same as Washoe, who had acquired 21 in the equivalent period. When she was three years three months, she had emitted 236 words, of which 78 met our criteria for acceptance.

By then Koko was regularly using such words as *love, hot, baby, time, necklace* (which she learned when we had to start using a leash on walks so she wouldn't dart into traffic), *small, blow, wiper* (meaning a cloth or paper towel), *pillow,* and *bread* (acquired when we started feeding her peanut butter sandwiches for non-meat protein). Her progress was heartening, not only because it compared favorably with Washoe's but also because it belied the gorilla's image as intellectually inferior.

Although Koko was constantly producing new surprises in her signing, it was when I reviewed her earlier signing performances that I was most struck by her increasing facility with the language. Our conversations six months into the project, when Koko was one-and-a-half, were definitely rudimentary: *

PENNY: Want *up?*

KOKO: *Up.* (I pick Koko up.)

PENNY: *Come* here, Koko.

(Koko comes over to me and we return to the nursery from the back storage area. I am holding a rubber man doll Koko wants.)

PENNY: This is a *man.* (I mold her hands to form the sign *man.*)

KOKO: *Food out more.*

PENNY: *Man.* (Again I mold *man.*)

KOKO: *Drink.*

PENNY: *Man.* (Again I mold the sign.)

KOKO: *Out.*

(Again I mold the sign. After a few more moldings I hide the man doll under my smock. Koko looks for the doll.)

PENNY: *Where* is the *man? Where?* (Koko brushes dirt off the bottom of my shoe.) *That dirty.*

By age two-and-one-half, Koko's signing was much more frequent and varied. On November 11, 1973, for instance, we had another conversation about going out. This one began with my spinning Koko around as she lay on the counter.

---

*In this and other conversations in the book, the human statements are made in both voice and sign language simultaneously, except where otherwise specified. As stated earlier, signed words are always indicated by italics. Thus, in the statement, "*Where* is the *man?*" all four words were spoken, while *where man* was simultaneously signed.

A hyphen between two signs indicates either that the two words were signed simultaneously (such as *go-there*) or that the sign translates to two different words in English (such as *frown-sad*).

KOKO: *Tickle.*
(I sign *tickle* on Koko's hand.)
PENNY: *What* do you *want?*
KOKO: *Out key.*
PENNY: *What?*
(Koko turns and looks out the window. I get out my keys.)
KOKO: *Open sweater key.*
(The sweater Koko wears on outings is kept in a locked cupboard. I hold up the keys.)
KOKO: *Key.*
(I give Koko the keys.)
KOKO: *Key key.* (She shakes the keys up and down.)
PENNY: *Koko plays with keys.* (As she plays, I bring some cottage cheese.) *Cheese for you. Give me* the *keys, Koko.*
(Koko hands me the keys, then pushes me around and climbs onto my back. I carry her around piggyback for a minute, then drop her off at the counter by the cottage cheese.)
PENNY: *Sit here.*
KOKO: *Out nut bean key.*
PENNY: *Cheese.*
KOKO: *Bean.*
(I mold the sign *cheese.*)
KOKO: *Open.*
(I again mold *cheese.*)
KOKO: *Bean.*
(I give up and give her some more cheese.)
KOKO: *More food.*
PENNY: *Want more?*
KOKO: *Out.*
(I mold the sign *cheese,* and offer her another spoonful.)
KOKO: *My cheese eat...food.*
PENNY: *More?*
KOKO: *More bean...white food.*

A year later, at age three-and-one-half, Koko still liked to go out, although by this time her signing had developed to the point where she could be much more explicit in her requests. On this occasion Koko's desire to go out was prompted by the appearance of our adopted cat, KC (for Koko's cat), at the window of the trailer we had recently moved into. I called, "Here, kitty,

kitty, kitty," and Koko, hearing this high-pitched chant for the first time, stared at me in apparent surprise, and then climbed onto my back to get a better look at the cat. Koko took my finger and put it on the door.

> KOKO: *Do key do key.*
> (I mold the sign *open*.)
> KOKO: *Open.*
> (I open the door and take her piggyback down the hall to turn down the heat. As I do so I mold *ride*.)
> KOKO (as we turn around to go back): *Go there.*

When we returned, Koko tore around the trailer for a minute until I caught her and brought her back to the kitchen. She went to her potty and signed, *Cat cat cat cat*. Then she returned to the window to look at the cat, who was in the grass hunting. She signed, *More there*, took my chin in her hand, pointed to my mouth, and signed, *More more there*. Wondering if she wanted me to repeat the call I made to KC earlier, I signed, *More cat say?* She replied, *Cat*. So I again called, "Here, kitty, kitty, kitty," to her apparent delight and satisfaction.

Koko's days at the zoo were not entirely occupied with language training. One memorable diversion was a party we had for Koko on her third birthday. The party began at 6:00 P.M., after Ron and I had spent an hour and a half preparing for the festivities. Naturally, the first thing Koko did was to open her presents. Barbara Hiller had brought Koko a 3-D viewer with animal pictures; Lee White, a volunteer, brought a wicker bed, a shrunken head, and a plastic snake that slithered down a stick; Ron gave Koko a quart-sized red glass; and I brought a volleyball, binoculars, a toy frog, rings, and a Snoopy piñata filled with nuts, candy, and toys. We hung the piñata from the ceiling of the trailer. Koko signed *look* when she took up the binoculars. She looked through them, and then tried to unscrew the eyecaps (Lee had some time earlier given her a fake pair of binoculars that converted into drinking flasks). Failing to detach the eyecaps, Koko put the binoculars around her neck and walked around like a field marshal.

The destruction of the piñata was a wild and wonderful event.

After knocking it down with one deft leap, Koko tackled it with hands, feet, and teeth. As the candy and nuts spilled out of a hole she made, Koko was overcome by the sudden deluge of such riches. She stuffed the treats into her mouth in a frenzy, eating candy wrapper and all. When miniature marshmallows fell out of the piñata, however, Koko became cautious and nibbled them in tiny bites.

Koko ate her birthday cake decorously with a spoon, but when she got to the last bite, she temporarily forgot her manners and scooped the cake directly off the plate with her mouth. We let Koko stay up late after the party. She was content to sit quietly in her new wicker bed hugging a stuffed gorilla toy as Ron and I ate our dinner.

6

The Move to Stanford

If I was elated at Koko's breakthroughs during the early days of the project, my enjoyment was tempered by the frustrations and crises of pursuing my work in front of gawking visitors. Although at first I had no idea how long the project was going to last, I still wanted to get Koko out of her glassed-in cage and into quarters that were more tranquil and private. Mr. Reuther gave permission for us to move Koko to a trailer, if we could find one that fit next to the gorilla grotto. So Mr. Reuther and I went trailer-shopping in San Jose. After visiting several dealers we found a used, partially furnished ten-by-fifty-foot mobile home in the want ads. It was over ten years old and a bit run down, but at $2,000 it was a bargain. In the fall of 1972 the trailer was installed safely out of view, next to the zoo's office trailer but unfortunately close to the tracks for the zoo's miniature steam engine. It took some time to adapt Koko to the trailer. Each day, if the weather permitted and the trailer wasn't being used for other animals, I would walk Koko from the nursery to the trailer to get her accustomed to it. If I was not there in the mornings, an assistant would accompany Koko. One assistant, who was somewhat overweight, occasionally showed up at the nursery perspiring heavily and sans gorilla. Koko, at first frightened by the new trailer, would escape and lead him on a merry chase back to the nursery.

Koko's gradual adaptation abruptly speeded up one day in June 1973 when she broke the glass window in the nursery kitchen area. A woman had knocked on the glass and Koko had knocked back a little too hard. Worried that Koko might repeat this performance, Mr. Reuther ordered her to move to the trailer full time, whether she was adapted or not.

This news seemed to me a fitting part of a miserable day. I had arrived at my office to discover that Koko had bitten her good friend Barbara Hiller on the hand. Then, in the mail I received word from a foundation that they had no funds for my project, along with a huge bill from Master Charge because of a computer error. On the way back to Stanford from the zoo I was looking forward to a relaxing dinner out with Ron, a respite from the day's tensions, when I received a speeding ticket. At dinner, the waitress spilled wine over my dress as she was about to serve the main course. My mood was not improved when I saw that the restaurant discriminated against women in the size of its portions. (I eat one *large* meal a day.) Finally when I got home I discovered the heater had malfunctioned in my pet iguana's cage and nearly roasted him alive.

Still, it was a good thing to move Koko to the trailer. At the nursery she had been learning not only language but also the basic skills of breaking and entering. Or rather, breaking and exiting, since it was a jailbreak that she had in mind. Even at her tender age, she had learned to work padlocks and twistlocks loose, and once she nearly got out the rear door of the nursery.

The trailer had a kitchen, an adjoining living room, and a hallway that led to a small bedroom, bathroom, and "master" bedroom. My assistants and I took turns staying in the large bedroom overnight when necessary. Like many small children, Koko began having nightmares after moving. She would scream, wake up, then fall back to sleep, or sometimes keep on shrieking once she awakened. When this happened, whoever was spending the night picked her up, comforted her, gave her some warm milk, and then put her back to bed.

After Koko became accustomed to her new trailer, we continued to take her for walks around the zoo. Occasionally we encountered a friendly mounted policeman. Koko was afraid of his horse, but she liked the policeman. One day the policeman mimicked the sound of a galloping horse for Koko's benefit. When we returned to the trailer I heard Koko imitating the clicking noises the policeman had been making. Since gorillas are not supposed to be able to imitate sounds at all, I was reluctant to believe my ears. Subsequently, though, Koko has imitated other unvoiced noises.

Because the trailer was carpeted, we stepped up our efforts to toilet-train Koko. By now we had had many more successful uses of the portable toilet than mistakes, and there was a pattern to the failures indicating that many of them might have been intentional. By July 1973, the great proportion of Koko's lapses occurred when she was locked up alone in the trailer at night, and were probably produced by the anxiety of being left alone. It is also remotely possible that Koko, noting our interest in her use of the toilet, figured that she might get us to stay with her by using it correctly only when accompanied by me or one of my assistants.

Koko's basic nature is fastidious. She has always hated stepping in dirt: outdoors she will insist that she be carried over puddles—if she can find someone to carry her—and indoors she will scrub and clean her quarters with a vigor that suggests more than mere imitation. Interestingly, the word *dirty*, which she first used at about age three, and which we use to refer to her feces, became one of Koko's favorite insults. Under extreme provocation she will combine *dirty* with *toilet* to make her meaning inescapable.

Even long after Koko had gotten used to sleeping alone, she still periodically failed to use her toilet, perhaps out of retribution or as an attempt at manipulation. On the other hand, gorillas, unlike chimps, do foul their nests in the wild, and so it is difficult to claim definitively that her nocturnal misadventures were manipulative or vindictive. Koko is, after all, unusually delicate on the subject of cleanliness, and it did not take long for her to become fully toilet-trained.

By the second year of Project Koko, my interest in Koko and that of the San Francisco Zoo had diverged to the point that some sort of conflict became inevitable. When I began the project it was with the understanding that at some point Koko would be reunited with the rest of the gorilla colony. The zoo felt a responsibility to try to breed and perpetuate this endangered species, and at first I accepted their logic that this could only occur if Koko was raised with other gorillas. Moreover, I shared the common belief that gorillas and chimps become unmanageable at about age six. My expectation was that I would work with Koko about as long as the Gardners had worked with Washoe—four or five years—and then return Koko to the gorilla grotto before she got out of hand. However, quite early in the project I began to wonder whether gorillas really do become unmanageable or whether environmental or other factors had made them appear so. Moreover, I began to wonder whether it was really necessary for Koko to go back to the gorilla grotto in order for her to have a baby. Many things contributed to these changing thoughts.

For one thing, I knew that a number of people had continued to work with adult gorillas. Carroll Soo Hoo had romped with full-grown gorillas; I thought if this man—who was smaller than I was—could get along with adult gorillas, then so could I. I had also visited the zoo in Basel, Switzerland, the December after the project began and saw a young male keeper in with several full-grown females and their offspring. He had no problem disciplining an adult and playing with the infants. And, while I did not disagree with the zoo's objective of breeding Koko, I thought it would be possible to breed her without returning her to the gorilla grotto. If we could provide her with another ape companion, I felt she could learn how to get along well enough with apes to mate.

Moreover, by now something more than cold objectivity was influencing my thoughts about Koko's future. Quite simply, she began to get to me. Koko was not just the subject of an experiment, she was a baby, and, I quickly discovered, as dependent and affectionate and engaging as any human infant. At first when Koko sensed I was about to leave, she would cling so fiercely that I literally had to pry her off before I could depart, and she sometimes left black and blue finger marks on my arms.

Caring for her entailed most of the joys and stresses of parenthood. And like a parent, I was endlessly fascinated by her development and charm. She cooperated with chores, assisting in cleaning and handing me items on request. She imitated my every move, from talking on the phone (Koko even opened and closed her mouth and huffed and screwed up her face) to grooming her fingernails when I did mine. She initiated hide-

and-seek games in which she would "hide" under a folding chair while I searched in cupboards and the oven, calling her name until finally she charged out laughing. Koko also continually sought and found trouble in various forms—dismantling her toilet, removing Formica from counters, setting off the timer on the stove, unraveling rolls of paper towels across several rooms, and feigning a hug while chewing up the tape-recorder microphone I wore attached to my smock. But any irritation would be dispelled when she'd wrestle with and kiss her dolls with loud smacks, tickle my ears, or make me a part of her bedtime nest by arranging my arms around her, gently pushing my head down into place, and lying down and cuddling.

As her vocabulary grew and Koko began to use words in ways that revealed her personality, I began to recognize sensitivities, strategies, humor, and stubbornness with which I could identify. It was the realization that I was dealing with an intelligent and sensitive individual that sealed my commitment to Koko's future. My knowledge of Koko's vulnerabilities made the prospect of returning her to the gorilla grotto unimaginable. By the time Koko was three, I was afraid that if that happened the trauma of separation would kill her.

Finally, I should also say that I was proud of Koko. The notion that another animal can acquire language is somewhat abstract until you see it happen or, in my case, make it happen. Then the world changes. My ambition to compare Koko's performance with Washoe's up to age four was only partly achieved, and so far Koko seemed to be matching—and in some ways exceeding—Washoe's performance. I desperately wanted to see how much Koko could learn, how far she would take her knowledge of language. But mostly, I wanted to continue to talk with her and be with her. The looming expiration point of my agreement with the zoo became an intolerable prospect.

The zoo had worried about the possibility that Koko might become too attached to me and humans in general ever to readapt to the grotto, but I doubt that they considered the possibility that I might become too attached to Koko to return her without a fight to what I believed was a potentially harmful situation. Actually, once it became clear that Koko was acquiring language, there was a division of opinion among the zoo officials and handlers about what would be the best future for the gorilla. Some worried that Koko was getting too humanized and would become unmanageable as an adult. Others realized that Koko was involved in something extraordinary, and several, principally zoologist Paul Maxwell, made efforts to help me continue my work unmolested.

However, my feelings about the life of a zoo animal were somewhat hardened by the experiences I shared with Koko behind bars. The zoo, under a new interim director following the departure of Mr. Reuther, decided to put Koko on display each day for at least a couple of hours because she was, after all, a zoo animal. And so for several months, Koko and I and whoever else was working with us spent daily periods on exhibit behind a chain-link fence. Koko did not seem to mind this much as long as we were with her. I, on the other hand, hated it. The fence let in all the raucous sounds of passers-by, and was not effective against the small objects that the more insensitive spectators would throw at us. I also worried about the danger of pneumonia presented by the sudden temperature change from Koko's heated trailer to the chilly and often foggy cage. At first I dressed Koko in a sweater, but officials wanted to put a stop to this on the ground that gorillas do not wear sweaters in the wild. This argument seemed absurd to me, since neither do wild gorillas spend their time locked up in cold, confining, prisonlike cages. After a few weeks in this cage, the glassed-in quarters of the nursery began to look quite cozy in comparison.

I began to try to think of alternatives that would satisfy the zoo's desires to breed Koko without terminating the language project and subjecting Koko to the stresses of zoo life. One of my ideas was to find a young chimpanzee as a temporary non-human companion for Koko. It seemed to me that it would be easier to find a chimp than another young gorilla, and that a chimp as a friend would be sufficient reminder to Koko during her formative years that she was an ape. This idea fell flat with the powers at the zoo. Then, in the fall of 1973, Paul Maxwell suggested that I get in touch with Marine World in Redwood City, which had a good-natured young male gorilla named Kong who was not much larger than Koko. I grasped at this suggestion as the only satisfactory alternative to Koko's reintroduction to the gorilla colony, and arranged for Koko and Kong to visit each other. Although the chemistry between Koko and

Kong never progressed to biology (both were much too young to breed), their brief liaison did serve the purpose of getting Koko out of the zoo and into her somewhat more tranquil quarters at Stanford. This was one positive thing that came out of that confusing period in the project.

At this point some zoo officials, already worried that Koko had become too estranged from her own kind to be reintroduced to the group, talked of "surplussing" Koko. This meant lending or selling her to another zoo. Once Kong was proposed as a companion for Koko, the idea of selling Koko to Marine World was batted about for a time. If nothing else, this indicates the uncertainty that clouded Koko's future.

Ironically, what most facilitated Koko's move to Stanford was that once it was agreed Kong was an appropriate companion for Koko, neither the zoo nor Marine World wanted its gorilla to make the long commute to the other facility. Marine World did not want Kong to visit Koko at the San Francisco Zoo because officials there were worried that their valuable and rare charge might pick up a stray infection and die. The zoo, on the other hand, worried that an auto accident might occur if Koko were on the road to Marine World every week. As a solution to this impasse I proposed to move the trailer to Stanford, where the danger of zoonomic diseases could be more effectively controlled, and where Koko would be only a ten-minute drive from Kong.

At this point I had invaluable assistance from several people. Richard Atkinson, then head of the Stanford Psychology Department, negotiated with Stanford to get permission and find a location on campus for Koko's trailer. Our lawyer, Edward Fitzsimmons, negotiated with the zoo for the purchase price, and Karl Pribram, my original advisor, contributed some of his grant money toward the buying of the trailer.

These negotiations were not without their amusing moments. We discussed several sites for the trailer. One spot we considered ideal was rejected, purportedly because a powerful administrator did not like the idea of a trailer spoiling his view of the campus. Eventually, we were given permission to locate the trailer in an area for laboratory animals. While to me the area had unpleasant associations with vivisection, it was relatively spacious and secluded.

A final logistical problem was to find funding to pay the zoo for their improvements in the trailer, and to pay the costs for operating Project Koko after we were installed. Once again, Richard Atkinson provided invaluable help. He and a biology professor, Donald Kennedy, who is now president of Stanford, lent their considerable reputations to obtain a grant from the Spencer Foundation that covered a large part of the costs of the project during its first two years at Stanford. Now all that remained was to convince Koko, the object of this ongoing custody battle, that the move was a good idea.

Moving day was September 19, 1974. The weather was foggy and somber. To lessen the trauma of the coming dislocation, I gave Koko four teaspoons of Benadryl. Unfortunately, Koko was so keyed up that this mild tranquilizer had no evident effect. While Ron and I waited for the workmen to prepare the trailer for the move, Koko chased passing peacocks. We had a hard time keeping her in one place. Our actual departure was rather solitary, reflecting the strained feelings that had surfaced during the dispute over Koko's future. Only John Alcarez, the gorilla keeper, came by to wish us good luck.

We left for Stanford at about 10:15 A.M., after the trailer was safely on its way. Ron, Koko, and I got in the car to begin the drive, and Koko, whose favorite pleasure is a drive in my car, happily signed *Go go*, and then as we continued around Lake Merced toward the freeway, *Go chase up*. After thirty minutes, however, Koko began to get anxious, reverting to her pre-toilet-trained ways and making the last part of the ride less pleasant than the first.

We arrived at Stanford well ahead of the trailer. To help abate Koko's mounting anxiety, we spread a tarpaulin in a shady spot near the University Museum, which adjoins the lab animal area. Koko decided that she had had enough of this outing, and signed *Go home*. She also signed *Go me Kate key* (Kate was her teacher, one of my assistants), perhaps to express her desire to return to the safety of her trailer.

When the trailer arrived an hour later, Koko was moved into another fit of expressiveness. She signed *Go home*, and then punctuated this statement by making repeated lunges for the trailer. It took some time to install the trailer, and so we had to restrain her. Once the trailer was ready for Koko to go inside, it still lacked electricity. Her anxiety increased when she heard the strange noises of buses and roosters and other activities that were part of her new surroundings. During that first night Koko awoke repeatedly and cried, and I stayed with her every night for the next month until she became sufficiently accustomed to the sounds of her new home.

Kong did not work out as a companion for Koko, mainly because his visits were not frequent enough for the two to form a relationship. We had expected that Kong would be brought to visit Koko at least once a week. It worked out that the two gorillas saw each other no more than once a month. One problem was that Kong was getting big and Marine World was having difficulty handling him. Moreover, he was not learning any tricks, a fact that comes as no surprise to anyone familiar with the gorilla's distaste for being told what to do. Eventually, Marine World offered to sell us Kong, but, acting on advice, we decided not to buy him. By the time Kong was offered to us in the spring of 1975 he was adolescent, and I felt that it would be difficult to come into his life that late and establish the dominance necessary to be able to handle him. Eventually, he was purchased by the Salt Lake City Zoo. Although Koko and Kong's liaison did not work out, it did get Koko and me to Stanford, where I could concentrate exclusively on the language project.

In the nursery of the San Francisco Children's Zoo, Koko, at age eleven months, feeds herself with a spoon as zoo-goers look on through a window.

Koko spent several months in the home of Deedee and Landis Bell recuperating from an illness. Here, at about age seven months, she plays with her first "pet," the Bells' cat Barney Google.

Koko at age one.

Koko, at about age two, signs *eat* for a spoonful of cottage cheese from Penny (above) and *listen* for her toy telephone (below).

Barbara Hiller, one of Koko's oldest friends, tries to pull a loose tooth.

Shortly before the Gorilla Foundation bought Koko from the San Francisco Zoo, her trailer was moved to the campus of Stanford University. Here, Penny and Koko squat on the steps of the trailer at its location in the lab animal enclosure.

Koko's room in the trailer, neater than usual. Out of view is a kitchen separated by a chain-link fence.

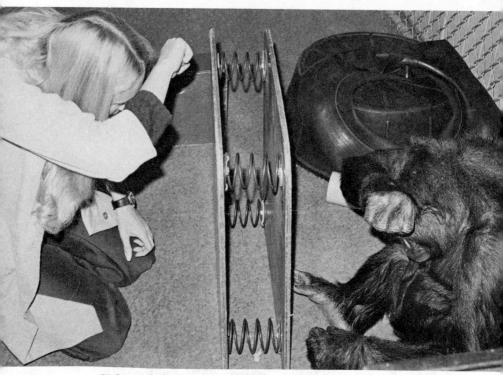

Koko and Penny sign *hide* in a game of hide-and-seek.

Penny gives Koko a lift so she can peek into a window of one of the Stanford buildings.

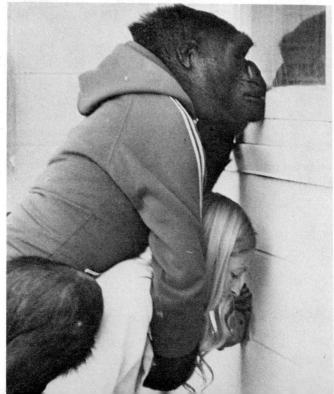

Koko with the object of her first schoolgirl crush, Al (whom she called *Foot*),
who did construction and laboratory work across from the trailer.

Penny, a bag of cookies in her lap, asks Koko, "Want?" Koko replies *cookie*. When Penny gets out a cookie, Koko signs in sequence:

Good ...

all ...

want ...

eat.

Penny gives Koko the
cookie she has earned.

As Penny tells the story of the three little kittens who lost their mittens, Koko comments that their mother is angry and the kittens are crying. Then she signs *bad*.

As a part of a vocabulary test, Penny asks Koko to find *crying* and Koko points to a picture of a child crying.

With an auditory keyboard linked to a computer, Koko produces
English words as well as signs.

Penny asks Koko *why* she wants the lights off, and Koko signs *sleep*.

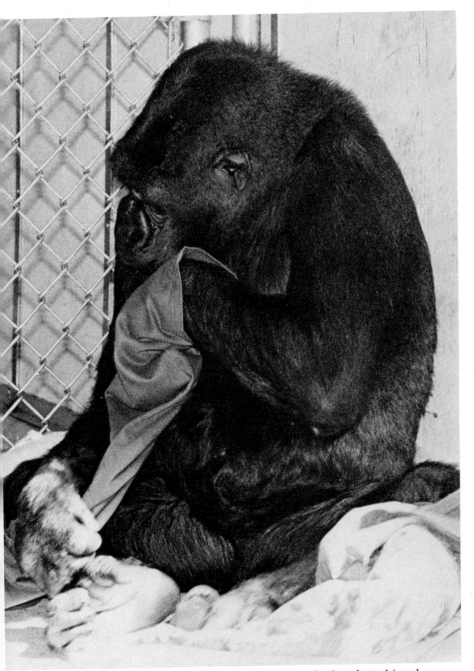

Making a nest with old clothes and rugs, Koko signs *soft*—hand stroking down the face—to herself.

Koko has always enjoyed having pets, whom she treats with gentle affection. Here, at about age five, she relaxes with her rabbit.

With our recent move to Woodside, Koko now has the freedom to climb a tree without attracting spectators.

7

The Campaign for Koko

Even after the move to Stanford, there still remained the question of who would have ultimate custody of Koko. Shortly after the move, Mr. Soo Hoo asked Ron whether we would be interested in purchasing Koko. Ron immediately said yes, but getting zoo officials to agree to this idea turned out to be a problem. During the negotiations over our move to Stanford, Saul Kitchener became the new head of the San Francisco Zoo. He was willing to let us have Koko, but only on the condition that we replace her with another female gorilla. This stipulation posed much more formidable problems than the purchase of Koko, no matter how steep the price turned out to be. Because gorillas are an endangered species, they are not—nor should they be—readily obtainable through animal dealers. Through various contacts, we approached a number of zoos and research centers. There was a seven-year-old female available at Yerkes Regional Primate Research Facility in Atlanta, but Mr. Kitchener turned her down because she was arthritic. Another gorilla was available from the Honolulu Zoo, but Mr. Kitchener felt that this one was too old. Then in 1976 Barbet Schroeder, the film director, put us in touch with an animal dealer in Vienna who was offering an infant female and an infant male gorilla for $28,000. When an animal dealer offers wild-born infants for sale, one can usually assume that the infants were "harvested" through the gruesome expedient of shooting the mother. In this case the dealer told us that he obtained the two gorillas in

Cameroon, and that they were orphaned after natives had eaten their parents. We were in no position to verify this story about the gorillas' provenance, and, ultimately, we overcame our qualms and decided to buy the two. The idea was that we could then give the female to the zoo as a replacement for Koko and keep the male to be Koko's eventual mate.

This left the simple matter of raising $28,000 to pay for the two infants. Together Ron and I had enough money to put a down payment on one gorilla. We still had a shortfall of about $21,000. At this point the media proved invaluable. Since the beginning of the project, my work with Koko had attracted a considerable amount of media interest. I would like to think that this attention derived solely from the awesome import of being able to converse with another animal, but I have had to accept that part of it centers on the supposed drama of a woman working with a "ferocious beast." In any event, during the period when we were trying to raise the money to buy the two baby gorillas, I would mention Koko's precarious future to the reporters who requested interviews. The local press took up my problems as a cause.

The two baby gorillas arrived on September 9, 1976. Their names were listed as King Kong and B.B. (short for Brigitte Bardot). We took it as our first obligation to rescue them from their unfortunate names. King Kong we renamed Michael, but we never got a chance to rename poor B.B. The rigors of her travels proved to be too much for her frail constitution, and in spite of our desperate efforts to nurse her back to health, she died of pneumonia within a month of her arrival.

On March 9, 1977, the *San Francisco Examiner* published an article about the uncertainties of Koko's future. The article reflected the sense of urgency I felt about Koko's future, and quoted me accurately as saying that I felt Koko would die if she were returned to the gorilla grotto. It also quoted Saul Kitchener as saying that he had never heard of anything like that happening. (I might point out in retrospect that there were no precedents involving language-using gorillas on which either of us could base our feelings, although I had read several accounts of apes dying or wasting away after abrupt separations from their mothers or caretakers.) The article generated more than $3,000 in donations toward the purchase of the two infants, and also spurred a "Save Koko" campaign, complete with bumper stickers, that ultimately got national attention. The recipient of the donations was the Gorilla Foundation, a nonprofit organization Ron Cohn and I, with the aid of Edward Fitzsimmons, founded with the idea that it would hold trust over Koko, protect her interests, and abet the study and preservation of gorillas in general.

This period was the low point of the project. The problems of obtaining the gorillas, the frustrations of trying to keep the object of a two-year search alive, and the grief of poor B.B.'s death almost immediately upon arrival took their toll on my concentration and health. Not only was I getting up several times a night to tend to a sick and dying gorilla, but also during this time I was told that I had either Hodgkin's disease or sarcoid and would need a biopsy from my lung. To quarantine Koko from disease, I had to shower and change after each visit to the dungeonlike infirmary where the animals were being kept. Perhaps the only good news to come out of this period was the discovery that I had sarcoid (a relatively benign disease) and not cancer.

At this time Mr. Kitchener was still inclined to insist that we find another replacement for Koko. However, popular sentiment had reached a point where the then mayor of San Francisco, the late George Moscone, became involved and insisted that the zoo allow us to buy Koko. Kitchener has since said that he would not have permitted the sale without this pressure "from above." Thus in the summer of 1977, nearly three years after the idea was first broached, we were permitted to purchase our "humanized" gorilla.

8

Koko's Day

Today Koko is still housed in her trailer but it has been moved to the secluded hills of Woodside, California. The structure has been fitted with a number of ingenious wire-mesh barriers and doors which permit my assistants and me to work with Koko without being in direct contact with her—Koko has been known to take advantage of her size when alone with some of the newer volunteers. Koko and Michael live within a few steps of my house, so that I can continue the routine of training established over the nine years of Project Koko. That routine consists of a daily mixture of language instruction, review, exercise, meals, and play for Koko and Michael, both separately and together.

I wake Koko up at 8:00 or 8:30 A.M. if she has not already been roused by Michael's antics. Following a breakfast of cereal or rice bread (rice and cereals plus raisins baked into a cake), milk, and fruit, Koko helps with the daily cleaning of her room. She enjoys going over both her room and Michael's with a sponge. Often these cleaning sessions end when Koko, seized by some urge, rips the sponge to shreds.

Then, some mornings, she sits on a chair before an electric teletype keyboard in the kitchen for a thirty-minute lesson in the production of English. Gorillas cannot generate the sounds necessary to speak, but through this Auditory Language Keyboard, which is linked to a voice synthesizer, we have given Koko a device that enables her to talk as well as generate signs. Other mornings we videotape or audiotape or work with flash cards.

These lessons generally end when Koko requests to have Michael in for a "visit." The morning play session lasts about an hour and is filled with tickling, tumbling, wrestling, chasing, and games of hide-and-seek. Only occasionally are there quiet moments during which the two catch their breath, or Koko grooms Michael.

At 10:30 Michael's teacher arrives and Michael returns to his part of the trailer. Koko, an assistant, and I chat in an unstructured manner for the next half-hour. Then, at 11:00, Koko has a banana and milk snack, following which sign language instruction starts. Her teacher introduces new concepts, reviews some of Koko's old vocabulary, and then acquaints Koko with some of the meanings of the "signs of the month." Lessons, spot quizzes, and tests are broken up by meals, snacks, games, and small talk. At 1:00 Koko has a light meal—a vegetable, meat, juice, and vitamin tablet. Meanwhile I get out materials for the afternoon—crayons and paper, magazines, books, and toys. More play and instruction follow. At 2:00 or 2:30 she gets a peanut butter and fruit sandwich. I return at 3:00 and invite Mike in for another play session, or, if the weather and Koko have both been good, take the gorillas outside for a walk.

At 3:30 Koko enjoys another snack, usually a cottage cheese and wheat germ mixture. At 4:30 she has a dinner of fresh vegetables. Her preferences start with corn on the cob, run through tomatoes, green peppers, cucumbers, sweet or white potatoes, green onions, peas or beans, squash, parsley, lettuce, and end with Swiss chard, spinach, and celery. Occasionally she samples artichokes, asparagus, eggplant, or other gourmet treats. Although she is open-minded about most new foods, she loathes olives, mushrooms, and radishes. Sometimes I dress up spurned vegetables with yogurt. Koko always has a glass of milk with her meal. If she cleans her plate, she gets dessert—either a cookie, Jell-O, dried fruit, or cheese and crackers.

After dinner Koko relaxes by leafing through a book, or nests with her blankets and dolls. Some evenings she asks to visit Michael's quarters. Koko especially enjoys romping in his training room and charging up and down the trailer hallway. After a bedtime ritual of toothbrushing—I brush her back teeth, she brushes the front—and moisturizing hands and feet with baby oil, both gorillas settle down at about 7:00 or 7:30.

Koko retires to a bed of three or four plush rugs placed over a large motorcycle tire. She has always been an accomplished nest builder, preferring her own creations to those designed by humans. Before settling on her current model, Koko has experimented with inner tubes, parts of her rubber toys, and other soft materials. I leave Koko with a night dish—a small fruit treat designed to make bedtime more pleasant. Even so, on some nights, Koko whimpers or gives her "whoo-whoo" cry when I leave.

PART II

Comparing Koko with Child and Chimp

9

The Rules of the Game

There is an enormous difference between a field study of animal behavior such as Jane Goodall's work with wild chimps in Tanzania and a "laboratory" study such as Project Koko. In a field study the scientist tries to minimize the effect of his presence in the animal's habitat and record all observations of whatever aspect of the animal's behavior is of interest. As Goodall and Dian Fossey, who has been involved in an extensive field study of gorillas, have demonstrated, a principal prerequisite for successful field studies is extraordinary patience and concentration. Project Koko also has required patience and dedication, but there the similarities to a field study end.

Koko was born in captivity and raised by humans. My responsibility was not merely to observe Koko but also to ensure her physical and emotional well-being. While the field study is concerned with what an animal does naturally in the wild, Project Koko was designed to teach a gorilla to do something no gorilla had done before. Project Koko required the insights into the gorilla personality that might come from a field study, but it also needed carefully planned training routines, as well as elaborate controls and testing procedures to ensure that whatever information it obtained was reliable.

If I was exhilarated with Koko's first words, I soon began to encounter the frustrations of trying to document something as elusive as language. I had hoped and expected that language would become a part of Koko's life and that she would use it for *her* own purposes as she learned about the sign language we were teaching her. Indeed, it would have been strange if Koko had responded to our promptings in precisely the forms we encouraged.

Instead, from the first months of the experiment Koko introduced her own variations and novelties into her signing. And while we were trying to prove empirically that Koko knew and used correctly the small number of words we had begun to teach her, it was her errors that gave us elusive glimpses of language games extraordinarily more sophisticated than the simple vocabulary growth we were looking for. And so almost from the beginning, as I developed procedures to map Koko's progress in a controlled and testable way that would produce credible data, I also tried to think of ways to capture and analyze those extra, teasing things Koko would do spontaneously. From the beginning, then, the language project with Koko was two experiments: the first a tightly controlled attempt to gather the facts about Koko's use of language, and the second a case study of Koko's use of the language we taught her. In the first we tried to get Koko to cooperate while we tested various aspects of her language use; in the second we were constantly trying to keep pace with Koko and understand what she was doing. However, it was only because of the rigor of our efforts in the first experiment that I might have any confidence in my guesses as to Koko's intent in our case study of her language behavior. And that rigor required an enormous amount of time, materials, and support apparatus.

By the third month of the program, my schedule had moved up from five hours a day, seven days a week, to eight hours a day, seven days a week. I was assisted by two native signers who worked with Koko two or three afternoons a week. Besides a daily diary in which everything we observed Koko say or do in our presence was noted, I maintained several different systems to monitor Koko's language use. Periodically, we would take a one-hour sample of all the signs and activities Koko was exposed to as well as all of her responses. This involved logging the statements of her human companions as well as keeping tabs on Koko's sign use. I maintained a daily sign checklist on which were noted all the signs Koko had made during the day, the combinations in which they occurred, the number of times she repeated each sign, and anything unusual that might have occurred during signing. As Project Koko developed, I instituted monthly filming sessions, videotaping, and tape recordings. The tape recordings consist of running commentary on conversations and activities with Koko; each month I record ten one-hour sessions and one eight-hour session.

Apart from these methods of logging Koko's progress, we regularly administered informal and formal tests of Koko's vocabulary comprehension, her understanding of the relationships between objects and words, and the standard infant intelligence tests, which involve tasks like putting a round peg into a round hole.

Most of the early years of Project Koko were devoted to studying such basics as vocabulary size, the ways in which Koko used her vocabulary—for instance, the frequency of signing, mean length of her statements, and favored words—and the ways in which she understood the words she was taught. In this manner Project Koko followed the broad division of the study of language into production (whether Koko could generate words) and comprehension (whether she could understand them).

Before any of this investigation could take place, I first had to set up some guidelines on recording signs so that I had a basis for agreeing that Koko had said what an observer thought she had said. Transcribing the samples required another set of rules so that an independent observer might examine a record of a sign sample and know not only what Koko said but also any peculiarities of context or of Koko's execution of the gesture. As one might imagine, this made transcription a time-consuming and laborious process. If Koko used a sign incorrectly, that was recorded; if Koko held a sign, that was noted as well. Similarly, if one of Koko's gestures was prompted, this was noted, as was the way in which it was prompted—"t" meant that the gesture was touch-prompted; "m" meant that it was molded; and "i," that Koko had imitated a gesture. These rules and many others allowed me to re-create the conditions in which Koko used a word, and the manner in which she executed a sign.

We had a similar set of rules covering the transcription of Koko's combinations of words. We defined an utterance as a string of signs terminated if Koko's hands returned to a resting position, or if she used her hands in some activity other than signing, or if she sought eye contact with her companion indicating that she expected a response, or if she was interrupted. While we recorded every word we saw her sign, for purposes of calculating the length of Koko's statements we would discard immediate repetitions of a word; *You there there chase* we'd transcribe as *You there chase.*

Given all these considerations, it takes about ten hours to complete the transcription of a one-hour audio-taped sign sample. If the sample is on videotape it takes five or even ten times longer. The videotapes, for which we raised the money by the sixteenth month of the project, serve as an excellent control for checking how accurately I am recording Koko's sayings. I am constantly surprised at how frequently I miss seeing signs that Koko is making. On occasion I will record one of Koko's statements as an error, and then a check of the videotape will reveal that I had missed the point of what Koko was talking about.

Actually, I have discovered that the videotape imposes its own constraints on the behavior of both the ape and the experimenter. After reading a charge that videotape samples of a chimp using Ameslan were riddled with imitations and interruptions, I began to wonder whether anxiety might be partly responsible. It is exceedingly difficult not to feel pressured by the presence of a camera. Most people feel as though they must "perform," and in my case, get the ape to perform. The result is that one is prone to talk more, to badger the subject with questions in an attempt to elicit signed responses. To check this, I have compared early videotapes of Koko's performance with recent samples in which I have made a conscious effort not to alter my behavior when the camera is running. (We have yet to devise a system in which the camera operates without either Koko's or my knowledge.) As I suspected, when I treated the videotaping as just an ordinary part of the day rather than a special event, Koko's behavior was more relaxed, and the number of her imitations and interruptions dropped dramatically.

One has to accept that no matter how tight the controls, a certain number of Koko's signs will be missed or misinterpreted. The observer is always at least partially distracted by the demands of recording what Koko says, whether writing or speaking into a tape recorder. Sometimes Koko sits on the far side of the room, or the view of her hands is partially blocked, or the observer is just tired. The virtue of having a number of different ways of sampling and recording Koko's sayings is that through cross-checking we can reduce such circumstantial errors.

Then, of course, there is the question of reliability. How do I know that my associates are not merely making up things that Koko says, or, for that matter, how does the reader know that I am not making up Koko's words, or reading words into coincidental but meaningless gestures? I chose the Gardners' criteria—spontaneous and appropriate use of a sign on fourteen consecutive days—for purposes of comparison with Washoe. Ultimately, the Gardners' criteria proved cumbersome and unsettling to Koko's motivation because of the regimentation and daily drill sessions they demand. It was one thing to drill a sign such as *clean* on fourteen consecutive days during the first months of the experiment; but as the project developed and Koko began to acquire language as though it were part of the natural course of things, the regimented procedures prescribed by the Gardners' criteria dissipated Koko's manifest desire to learn.

In Koko's case, as in the case of the chimps, after an initial, intensive period of training, a phenomenon called "learning to learn" takes over. The animal seems to grasp what all the molding and repetition is about and begins to acquire new signs and concepts more and more rapidly. However, if Koko is persistently drilled on a sign like *giraffe* after she has learned it to her own satisfaction, she becomes restless, moody, and intransigent. Often she'll walk away. Or occasionally she'll make every possible variant on the word but the correct sign. *Berry*, for instance, is made by lightly grasping the thumb with the fingers of the other hand, and then pulling apart. Asked the sign for *berry* during one drill session, Koko grasped her index finger, middle finger, fourth finger, and little finger. Moreover, the opportunity to use many of the words she knows simply does not often arise. She will use a sign such as *balloon* correctly, for instance, but only once every few months does the occasion call for it. Koko knows the sign—she once used it appropriately and spontaneously without having seen a balloon for six months—and yet according to rigid formal criteria the sign is not part of her vocabulary. One month she used the word *cucumber* on thirteen consecutive days, but then not again for a couple of days. Thus, *cucumber* failed to qualify by the Gardners' criteria for reliability.

Given these considerations, I have developed my own criteria for Koko (although I have rigidly adhered to the Gardners' standards in all comparisons with Washoe): for a sign to be considered reliable it must be recorded by two independent observers and must be used spontaneously and appropriately on at least half the days of a month. Thus, a sign reported fourteen times, but by only one observer, would not be considered a reliable part of Koko's vocabulary. Clearly, this still requires extensive drill—and Koko recently gave her opinion of such sessions with gorilla candor. After drilling Koko on parts of the body, one assistant asked what she thought was boring. Koko replied, *Think eye ear eye nose boring.*

We developed other controls for the reliability of reports, to ensure that the data would not be skewed by erroneous reporting by one individual and then have the error compounded by transcription by the same person. One technique was the simultaneous recording of all of Koko's utterances by two people present at the scene. Another was the independent transcription of a videotaped session by two assistants familiar with Koko's signing. A third was the independent comparison of simultaneous audio and video tape samples.

With these controls and more, I hoped to have sufficient safeguards to still any doubts that Koko really does say what we report she says. At least any debate might then center on the meaning of what she says rather than on whether Koko actually says something.

These very problems associated with establishing criteria and testing procedures for Koko are what first spurred me to think about the merits of a case-study approach to Koko's language use. Instead of discarding all of Koko's variations as errors, I began to think that it might be more productive to re-examine some of these "mistakes," taking into account the gorilla's personality, the context in which she signed them, and her past signing record to see whether some of these variations might be intentional.

Drink is one of the first signs Koko learned. It was made with the hand in a hitchhiking position but with the thumb touching the mouth. She has used it thousands of times, and yet one day she persistently refused to make the sign when Barbara Hiller, one of Koko's oldest friends, requested that she do so. Barbara tried everything she could think of, but Koko would only sign *sip, thirsty sip,* and *apple sip*—anything but *drink.* Finally at the end of the day, Barbara said with some desperation, "Koko, please please sign *drink* for me." Koko leaned back against the counter and, grinning, executed the sign perfectly—to her ear. Koko makes equally elaborate "mistakes" without the grin, which indicates that her intent is not to joke but to be ornery. During a videotaping session in which we were reviewing prepositions, I asked Koko to place a toy animal *under* a bag. Deadpan, she took the toy and stretched it up to the ceiling.

On occasion Koko will go to extraordinary lengths to be contrary, and this permits the subtle human to almost program her actions. Ron got Koko to stop breaking plastic spoons by saying, "Good, break them," whereupon Koko stopped bending them and started kissing them. Koko is aware when she is misbehaving. Once when I chastised Koko for her bad behavior, Koko described herself as a *stubborn devil.*

Koko's contrary responses raise the question of how the disinterested observer should interpret data. In an ideal scientific world, the scientist seeking to compare Koko with a chimp or a child in the use of language would pull the data on each, examine it to make sure that controls and collection procedures make it comparable, and then proceed with an analysis. Life, of course, is rarely so straightforward, and never is life less simple than when one sets about to make judgments concerning the nature of language.

One of the criticisms leveled against the chimp experiments from various quarters was that they were not learning language at all, but were in fact demonstrating the "Clever Hans" phenomenon. Clever Hans was a wonder horse in Germany who in the late eighteenth century confounded everybody, including his owner, with his ability to do math problems. People tried all sorts of ways to ensure that the owner was not wittingly or unwittingly supplying the proper answer. Just at the point when the last skeptic was silenced, someone had the idea of seeing whether Hans could solve the problems blindfolded. The owner, not understanding the origin of Hans's genius any better than the spectators, readily agreed. Blindfolded, Hans was a dolt. It turned out that when presented with a problem, Hans would just start tapping his hoof, all the while keeping a sharp eye on his owner. The owner would innocently and almost imperceptibly straighten up when Hans reached the proper answer, and Hans would then stop tapping. Hans was clever, but he was no mathematician. And so the criticism of Washoe and other chimps was that what they do may look like language, but their teachers, however sincere they are, must be giving them inadvertent cues.

To ensure against inadvertent cueing, I have checked my findings through double-blind testing, in which the ape can see the test object to be identified, but not the tester, and the tester can see the ape's response, but not the object. For instance, I'll put a toothbrush into a plywood box with a Plexiglas front, cover the box, and then leave the room. Koko will enter from another room and sit down in front of the box. Standing behind the box, unable to see its contents, is an assistant who asks Koko, "*What do you see in the box?*" or "*What's that?*" and writes down her response. Koko then leaves, I return with another object, and we repeat the procedure. At no time does Koko see me, or the assistant see the object. This eliminates any possibility of cueing, and random changes in the order of the objects presented for identification prevent the ape from using a strategy like memorization to come up with the correct answers. Thus I can be reasonably sure that when Koko makes a sign she is referring to the object presented for identification. This rigor breaks down when one turns to comparative data on children. The Gardners once made the ironic point that if one is going to judge whether a creature has language based upon the rigor of the data collection methods, one could make the case that chimps and gorillas have language and children do not. An examination of studies of children's language acquisition reveals that these are shot through with examples that, if generated by chimps or gorillas, would have been rejected as prompted or unclear. It turns out that the methods for gathering data from children are open to the charge that they are influenced by "Clever Hans" considerations. The Gardners point out numerous instances of prompting that have occurred in the twenty years linguists have been gathering data on children. It is ironic that the collection of "hard" data on language development in apes has spurred the search for better controls when studying language development in children.

It might seem appalling that sloppiness has been tolerated in data collection on children, but it is easily explainable, and the explanation underscores the difficulties of determining what is a fact in a field as nebulous as language. The problem is that animal and human communication have been examined from opposite points of view. We have studied human communication with the foreknowledge that all normal humans eventually learn spoken language. And so we have tended to find the elements of adult speech in the child's earliest utterances, whether they are there or not. On the other hand, animal communication has traditionally been studied with the expectation that the animal will never master language. Thus the data produced by the two kinds of studies is mixed up with the assumptions that governed the collection of the data in the first place. Until recently this meant that the study of animal and human communication only reinforced the assumptions upon which the studies had been based.

There is much about language that does not lend itself to reduction to statistics and hard data, and some linguists have recently reacted against the rigid, formalized treatment of language. When we speak with each other, we are not isolated by double-blind procedures. Indeed, a good deal of our comprehension of the spoken message comes from a perfectly natural "Clever Hans" appreciation of the nonlinguistic cues to meaning of the message. This is not to justify vagueness but to illustrate that it is very difficult to speak with any confidence of "facts" about language.

One point that should be made is that critics miss a fundamental aspect of ape sign-language performance in dismissing it as a Clever Hans phenomenon. The horse Clever Hans had merely to look for cues that would tell him whether or not to continue tapping his foot: a go, no-go decision. Koko's options are hundreds of signs or thousands of sign combinations, and she frequently violates our expectations as to what her response will be to a particular question. Once I asked her how she slept, expecting an answer like *fine*. Instead, Koko signed, *Blanket there,* pointing to the floor. Anthropologist Suzanne Chevalier-Skolnikoff supports the conclusion that Clever Hans is not the most conservative or economical explanation of ape language use. "Apes manifest advanced cognitive processes non-linguistically," she writes, "and since they appear to manifest them in signing, it is illogical to attribute their signing to simpler cueing and Clever Hans."

When Koko made the *drink* sign in her ear, it would seem that she made an error. After all, she did not make the sign correctly. Yet, just as a standard test may be incapable of measuring the abilities of the gifted but unmotivated child, a strict interpretation of Koko's response would suggest that Koko did not know what she was talking about and was merely randomly generating gestures. However, the context makes it clear that Koko knew what she was doing, but decided to be uncooperative. Countless other examples of similar negativity underscore her intent on such occasions.

For instance, on January 1, 1978, I wanted Koko to sign *shell.* Out loud we asked Koko to sign *shell,* first showing her a shell. There was no response. "Forgot?" I asked. Still no response. Finally I sent Koko to her room and closed but did not lock the door. As I did so I said, "Well, I'll just take these goodies to Michael." At this point, Koko edged out of the door and, unprompted, signed *shell.* That same evening I wanted Koko to demonstrate the sign for *rock.* This time I did not have a rock to show Koko and so I tried to elicit the sign by saying in English, "What is the sign for 'rock'?" Koko made a number of bizarre gestures with her two fists but not the correct sign for *rock,* which is made by hitting the fist of one hand onto the back of the other. Finally, I said, "I won't give you your night dish unless you say 'rock.'" *Rock,* signed Koko.

Koko's stubbornness is an interesting phenomenon, because it also surfaces in humor and in a type of gestural cartoon where, as in the case of stretching to place a toy "on" the ceiling instead of "under" a bag, she does the opposite of what she is asked but in such an exaggerated way that she makes it clear she knows what she is doing. And it shows up in a type of verbal playfulness. Barbara Hiller encountered it one afternoon when she noticed Koko playing by herself. Koko was making a nest out of white towels, and as she arranged the towels, Barbara noticed that Koko was signing *red*. Barbara said, "You know better, Koko. What color is it?" Koko insisted that it was red, signing *red* three times, each sign larger than the preceding. Then, with a grin, she picked up a minute speck of red lint that had been clinging to the towel and held it up to Barbara's face, signing *red*. A few days later she went through exactly the same routine with me.

If these anecdotes reflect Koko's intentions, then she is turning her understanding of language to her purposes in quite sophisticated ways. Humor is an exceedingly complex phenomenon. It presumes an understanding of certain norms, which are then distorted in recognizable but preposterous ways. In making the sign *drink* in her ear, Koko was playing with the underlying structure of Ameslan, rearranging the constituents of the sign just enough so that she could send a far more complex message than the simple answer Barbara was looking for. Koko's response was a joke on Barbara, but it was also a joke on language, and in this sense it was metalinguistic—it used language to comment on language. Similarly, Koko's insistence on signing *red* could be interpreted as a joking comment on the monotonous literalness constantly demanded of her by her signing companions. The speck of lint was literally red, but Koko stretched the literalness to the point of absurdity given the sea of white that surrounded the red. It is the type of joke that might be made by any creature with a lively intelligence who was exasperated by the simple-minded, repetitious tasks it was constantly given. Keep in mind that the impetus for this creativity—Koko's moodiness—is a characteristic of gorillas.

This gulf between the complexity of what Koko regularly does with language as evidenced through anecdote, and what we can empirically demonstrate she does through controlled experiment, reveals the frustration of trying to document something as elusive and uncontrollable as language. While we have to call *drink* signed to the ear an error, then, we know it is, more importantly, a joke. We know this because it is characteristic of the way she expresses her contrariness, because we know how well she knows the sign *drink*, because we know of other times in which she has made anatomical jokes, because she was grinning, and because a host of corroborating circumstances surrounded the incident. However, to construct a study designed to document a phenomenon such as humor, in which interpretation depends upon the subjective intentions of the animal, would be an enormously complex and perhaps impossible task. Instead, we interpret the meaning of this anecdote through devices that are an ordinary part of understanding any message, but that do not fall neatly into the empirical method of any scientific discipline.

And so as Project Koko proceeded, we developed a picture of Koko on two levels. The first derived from strict interpretations of the hard data that has been collected. The second was based on an informed look at the anecdotal "case-study" evidence concerning her understanding of some of the more elusive aspects of language. In effect, we were watching Koko learn to ride a tricycle while she simultaneously turned in a creditable performance at the Indianapolis 500.

Before getting to the Indianapolis 500, however, it must be established that Koko can ride a tricycle, because it is only after we have proved Koko's mastery of a basic vocabulary that her innovations gain credibility.

10

Production: The Basics

In linguistics, production refers to the ability to utter words and sentences. In organizing my investigation of Koko's language use I followed the traditional approach of studying first her ability to *produce* words and then her ability to *comprehend* them.

My assistants and I have logged virtually every word we have seen Koko utter. Since with every month she uses more and more words, just keeping my records up to date is a formidable task. However, it is only as this mass of raw data is analyzed that I have been able to come up with a picture of Koko's language development, and in turn to make informed guesses about the more sophisticated usages Koko has demonstrated.

As the data began to accumulate, I analyzed Koko's word use in a number of different ways. Over the first five and one half years of the project, Koko's greatest spurt in vocabulary growth occurred between age two-and-one-half and four-and-one-half (see Figure 1). In human infants, the same spurt comes between ages two and four. Even after her surge in vocabulary growth, Koko has continued to acquire signs at a steady pace, but many do not qualify by our rigid criteria. Because she loathed the daily drills that were necessary to meet the criteria, I felt that once Koko's qualified vocabulary reached 200 signs, the basic point—that Koko could reliably acquire signs—had been made. We could then use these bothersome drills in the investigation of new phenomena.

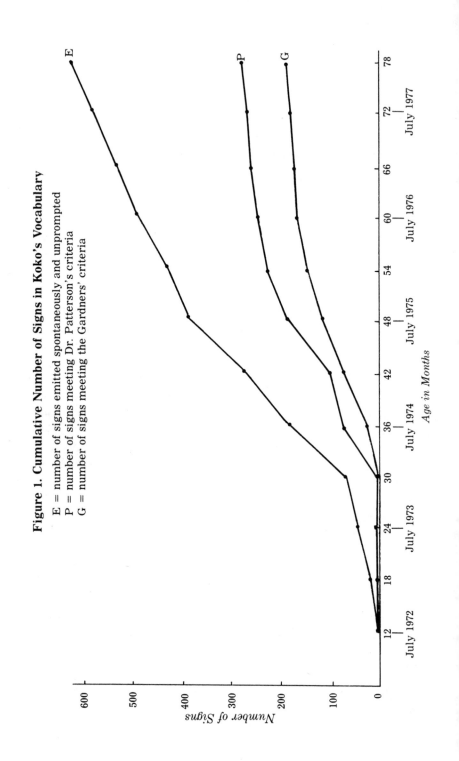

Figure 1. Cumulative Number of Signs in Koko's Vocabulary

E = number of signs emitted spontaneously and unprompted
P = number of signs meeting Dr. Patterson's criteria
G = number of signs meeting the Gardners' criteria

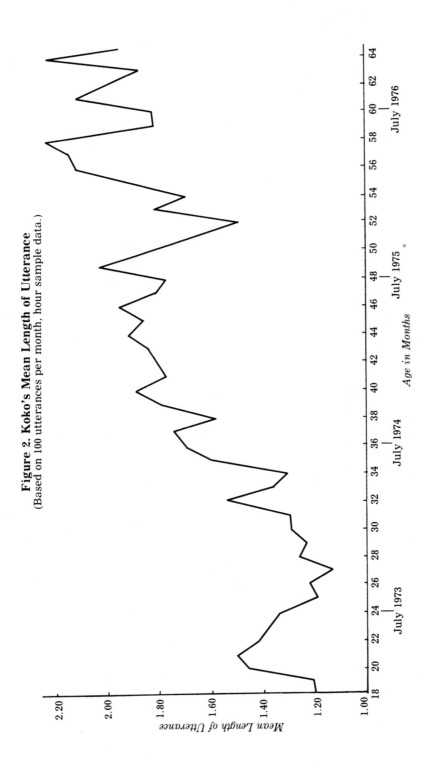

Figure 2. Koko's Mean Length of Utterance
(Based on 100 utterances per month, hour sample data.)

In the number of words used in Koko's utterances, another
great spurt came during that same age period of two-and-one-
half to four-and-one-half (see Figure 2). The variety of Koko's
statements also took a jump during this period (and for that
matter has continued right up to the present). Some typical
statements:

July 1972
KOKO: *Up.*

July 20, 1973
PENNY: Do you *want more?* (for more orange)
KOKO: *More-food.*

July 20, 1974
KOKO: *Tickle me Penny.*

July 20, 1975
KOKO: *More cereal me eat.*

July 20, 1976
KOKO: *Please hurry on necklace* (referring to the leash).

July 20, 1977
KOKO: *Stamp eat Koko taste* (referring to a gummed label,
   which Koko then eats).
KOKO: *Drink orange Koko thirsty drink.*

July 20, 1978
KOKO: *Love eat nut cracker sandwich* (for crackers with peanut
   butter and jam).
KOKO: *Love bread nice breakfast* (for breakfast rice bread—
   Koko uses *nice* to say *rice,* for which she was given no sign).
PENNY: *Who's that?* (I point to Ron.)
KOKO: *Visit devil trouble Koko that.*

July 21, 1979
KOKO: *Me love happy Koko there* (to photographs of her birthday
   party).

July 30, 1979
MAUREEN SHEEHAN: Can you say a long sentence about lunch?
KOKO: *Love lunch eat taste it meat.*

At age four-and-one-half the frequency of Koko's signing and the length of her statements waned, probably reflecting her sensitivity to changes in her life. In February 1976, at age four-and-one-half, Koko bit one of her deaf instructors, Kate Mann, who was trying to discipline her. Kate felt that she had lost the necessary dominance to work directly with Koko, and from this point onward worked with her separated by the chain-link barrier that seals the trailer kitchen and foyer from Koko's quarters. With the chain link between them, Kate could not keep Koko's attention from wandering as she could when they worked together, and Koko's signing dropped somewhat. Six months later Kate left to attend college and was replaced by another deaf assistant, Cathy Ransom. Koko missed Kate, and, as is the case with children, may have felt that she was abandoned because she was bad. Maureen Sheehan, another assistant, reports that she asked Koko about another of her favorites who had left. Maureen said, "Remember Cindy? How did you feel when she left?" Koko responded: *Back, Koko good.*

In the month immediately following Kate's departure, the average number of different signs Koko used daily dropped from 130 to 120. On two other disruptive occasions—when Koko became ill and when she was introduced to Michael—her sign production also dropped.

As the evidence began to accumulate, we were able to make preliminary comparisons with the data gathered on Washoe and see how Koko's progress measured up to a chimp's. Since the Gardners began their work, a number of experiments have attempted to teach chimpanzees a variety of languages. Still, the only experiment involving Ameslan in which a chimp's vocabulary was allowed to grow in a relatively unrestricted manner was Project Washoe.

Comparing Koko with Washoe according to even the Gardners' own criteria, Koko seemed to acquire signs at a faster pace. After three years of training, Washoe's qualified vocabulary was 85 signs, while Koko's was 127. (At that point the two had 46 signs in common.) At fifty-one months, Washoe was up to 132 qualified signs, while Koko had 161. By this point the chimp and the gorilla had 72 signs in common (see Table 1). Actually, the gap between Washoe and Koko is slightly smaller than these figures might indicate, because I included as indi-

Table 1. Comparison of Koko's and Washoe's Qualified Vocabulary (Gardners' Criteria) at Age 51 Months

KOKO: (89 signs)		
all	dry	orange
alligator	ear	peach
apple	earring	Penny
arm	egg	pepper
around	elephant	pick-groom
ask	eye	pig
bad	feather	pillow
bean	finished	pinch-skin
bellybutton	fish	pink-shame
belt	frown	potato
big	giraffe	pour
blanket	gorilla	ring
blow	grape	rubber
bone	hair	sandwich
bottle	help-myself	scratch
bottom	Koko	sip
bracelet	leg	skunk-stink
butter	lip	small
cabbage	lipstick	soap
cake	match	sock
candy	medicine	spice
carrot	milk	sponge
chase	monkey	straw
clown	mouth	sweater
cookie	nail	tape
corn	nailclipper	taste
cracker	necklace	teeth-glass
do	nose	tiger
don't	on	whistle
drapes	onion	

SIGNS IN COMMON: (72 signs)		
baby	comb	hurry
bag-purse	come-gimme	key
banana	cow	kiss
berry	different	knife-cut
bird	dirty	light
bite	drink	listen
book	eat-food	look
brush	flower	me
bug	fork	meat
cat	fruit	mine
catch	go	mirror
cereal	good	more
cheese	grass	nut
clean	hat	open
clothes	hug-love	out
cold	hungry-want	pen-write

Table 1.

	please	sorry	time
	quiet	spoon	toothbrush
	red	stamp	tree
	ride	string	up
	same	sweet	water
	sit-chair	this-that-there	white
	sleep-bed	thirsty-swallow	wiper
	smile	tickle	you

WASHOE:	airplane	Greg	pants
(60 signs)	bath	hammer	pin
	black	hand	pipe
	butterfly	help	Roger
	can't	hole	Ron
	car	hose	run
	climb	hot	shoes
	cover	house	smell
	cry	hurt	smoke
	cucumber	in	spin
	Dennis	Larry	Susan
	dog	leaf	telephone
	Don	Linn	tomato
	down	lock	Washoe
	Dr. Gardner	lollipop	we
	enough	man	Wende
	floor	Mrs. Gardner	who
	funny	Naomi	window
	goodbye	no	woman
	green	oil	yours

vidual signs certain body parts that are indicated simply by pointing, but the Gardners classified these 8 signs as *there*. Before one could assert that the gorilla is better at acquiring language than the chimp, one would need more data on several different chimps and gorillas. The difference in the rates at which Koko and Washoe acquired signs may not be significant.

As one would expect, Koko acquired language more slowly than a normal speaking child. Speaking human children often have vocabularies of over a thousand words by age three. Some of this discrepancy reflects the difference between sign language and spoken language. A deaf person, no matter how intelligent, has to make do with a smaller gestural vocabulary than that permitted by speech. (Finger-spelling resolves this discrepancy by permitting the deaf to encode any spoken word visually.) For instance, hearing children may acquire six hundred new words a year between ages two and six. As yet there is no clear comparable data on the average rate of sign acquisition for deaf children. The differences in the home situation (whether or not the parents know sign language) may account for huge variations among deaf children in the rate of sign acquisition.

A fluent signer can get by with as few as five hundred to a thousand signs, supplemented by finger-spelling, pantomime, and variants in modulation such as facial expression, body posture, and the manner in which a sign is made. Koko, too, supplements her vocabulary through modulations, although it is physically impossible for her to articulate about one-third of the manual alphabet.

Fluent signers also tend to use relatively small working vocabularies even when their vocabularies are quite large. It takes perhaps twice as long to complete a gesture as it does to say an equivalent word. This places a premium on economy of expression, and even a sophisticated signer might prefer to say *Shop me* rather than *I am going to the shop.*

One constraint on vocabulary size is, of course, the number of signs to which an individual is exposed. In general, mothers use a smaller vocabulary when they speak to young children than they do when they speak with adults. In my case, my sign vocabulary was restricted at the outset of Project Koko simply because at that point I did not know very many signs. This was not the case with the deaf assistants, who worked with Koko from the beginning of the project, although they too used a very restricted vocabulary at first. In five random one-hour samples taken in the summer of 1973, Koko was only exposed to about 80 different signs. Two years later, the number had risen to 189 signs, and by 1977, to 353 signs. This suggests, at the least, that Koko was assimilating a high percentage of those signs to which she was exposed.

As is true of hearing children, deaf children, and Washoe, slightly over half of Koko's early vocabulary consisted of names and nouns (nominals). By the time Koko was seven, nominals made up nearly two-thirds of her vocabulary—again, similar to the development of hearing children. Some characteristics of

Koko's vocabulary development more closely resembled those of deaf children than of hearing children. Like deaf children, Koko rarely used definite and indefinite articles, or the conjunction *and*, undoubtedly because of differences between Ameslan and spoken language. In Ameslan, *and* may be indicated by pointing in succession at whatever is being conjoined. Similarly, the deaf may say *this* and *that* by pointing.

Modal or auxiliary verbs like *be, have, can, could, might, must,* and *will* are also absent from samples of early deaf vocabularies I examined. The children only used *do* in the form of *don't.* *Have* and *do* were the only modal words present in Koko's vocabulary up until 1977. *Don't* and *can't* are now also in her vocabulary.

Koko and deaf children also use relatively few temporal words. In one sample of deaf children, only six temporal words occurred, with the most popular being *finished*. By now Koko has acquired a few words related to the concept of time—*now, time, finished*—and she uses the compound *Good finished* to mean "later" in precisely the way that has been reported for deaf children.

Koko has an excellent sense of time. She knows when people are supposed to come and go, and if you promise her something later, at the appropriate moment she will remind you of your promise. Maureen Sheehan has a regular schedule with Koko that changes every few months. Once, only four visits after Maureen's new schedule was established, I was delayed relieving her and Maureen had to stay an extra forty-five minutes with Koko. When Koko saw that Maureen was not leaving at her normal time, she looked at Maureen and signed, *Time bye you.* Maureen responded, "What?" and Koko signed, *Time bye good bye.*

Both hearing and deaf children seem to pick up prepositions slowly, using only two or three by age three. By the time a normal child is six, however, prepositions make up roughly one out of twenty words in the child's vocabulary. At age six, Koko knew a number of prepositions—*on, in, out, down, up,* and *around*—which amounted to 2.2 percent of her vocabulary.

Like all types of children Koko learned *no* before she learned *yes.* Currently she uses *yes* more often than *no,* but she does not use either sign frequently.

One aspect of Koko's vocabulary that distinguishes her from both deaf and hearing children is her relative lack and infrequent use of the interrogative signs. She responds appropriately to questions using *who, what,* or *why,* but she rarely uses the words in asking her questions. No question sign has yet met either my or the Gardners' criteria, although Koko has used several of these words spontaneously and appropriately on different occasions. Just before her seventh birthday, she signed *What?* when told that she could not come out of her room because there was a problem. She may have been asking what the problem was. Another time, when told *Open your mouth and close your eyes,* a suspicious Koko was moved to sign, *Why?*

Koko most commonly forms questions through a gestural intonation. She will hold the hands in the sign position and seek eye contact, which both deaf children and hearing children also do in the early stages of language acquisition, especially when asking "yes-no" questions. Hearing children are known to ask yes-no questions by using rising intonation or by looking at the listener after speaking, and deaf people ask yes-no questions by raising their eyebrows with an expectant expression on their faces and making eye contact.

This may seem like a vague way of forming a question, but Koko's questions are quite distinct from her simple declarative statements. For instance, when Barbara Hiller heard a woodpecker tapping outside the trailer, she signed, *Koko, listen bird.* Koko responded, *Bird?,* making the word a question by holding the sign and turning toward Barbara, cocking her head, and slightly raising her eyebrows. Barbara signed, Yes, *that's a bird* tapping *outside. Listen.* Barbara then tapped to imitate the sound of the woodpecker and pointed outside. This time Koko signed, *Listen bird,* with a declarative intonation.

It is possible that Koko rarely uses question signs not only because we rarely teach them by molding, but also because there have been few pressures on her to be more specific in the forms of her questions, and because she has been adequately served by her inflected form of interrogative. The humans who work with Koko are constantly asking her all sorts of questions, and are also very responsive to anything Koko says. Barbara interpreted Koko's question, which was phrased with a great economy of motion, and then gave her an elaborate answer complete with pantomime. If you are getting that kind of service with eye contact and a simple cocking of the head, why bother to change?

When children begin to learn language, they also begin to learn about language. As they learn a word, they will develop rules about the situations in which the word applies based upon the circumstances in which they learned the word. They will refine their understanding of the word by applying the word to a host of situations—by playing with language. Parents will correct children's inappropriate applications of the word, and thus children gradually develop a sense of which properties of a given word are general and which are specific.

The way in which a child overgeneralizes a word seems to depend on a number of factors. Some children consistently extend the meaning of a word on the basis of their perceptions of the referents, while others are not consistent. The parents affect a child's overgeneralizations by their reactions to the way a word is used; the length of the interval between the acquisition of a word and its extended use also conditions the child's overgeneralizations.

Overgeneralizations are most common in children between the ages of one and two-and-one-half. Every man becomes "dada"; every animal "cat" or "dog." Koko overgeneralized for a more extended period of time. Early in the project, some of Koko's overgeneralizations were awesome in their scope. She acquired the combination *eat-food* early, and at first she would apply it to any object, edible or inedible, that she wanted to put in her mouth. When Koko learned *bean*, she clung to the word with a vengeance, substituting it for more appropriate words in her vocabulary. She seemed to find the sign fun to execute— grasping the tip of the index finger with the thumb and index finger of the other hand and pulling, like removing a glove from one finger. She used *bean* for anything from cookies and shoes to artichokes, toy tigers, Jell-O, and a person with Koko's toy glasses propped on his head!

Koko's later overgeneralizations show the powers of analysis she brought to bear on her world. For example, she learned the word *straw* at three years two months to mean a drinking straw. She then spontaneously used the sign to label plastic tubing, a clear plastic hose, cigarettes, a pen, and a car radio antenna. In the absence of other words for these objects her conclusion that *straw* referred to long thin objects was logical if mistaken. *Tree*, which she learned for acacia trees and celery stalks, she also used to refer to scallions, asparagus, and a variety of tall, thin, vertical objects.

Koko also overgeneralized on the basis of color—although, as with children, this was rare. Apparently operating on the assumption that *grass* was synonymous with *green*, Koko called a green pig a *grass pig*, a head of lettuce, *lettuce grass*, and both a green apple and green toy mouse *grass*.

Like children, Koko also overextended words that relate to a similar function. She confused the signs *sponge* and *soap*, and also the signs *meat* and *bone*.

Early in the project she also seemed to overgeneralize on the basis of the form of a sign rather than on similarities of its referents. It is akin to a child who uses the word "barbecue" to refer to a barber shop, or "pajamas" to refer to a piano. Koko would occasionally sign *clean* to ask to be allowed out or to be lifted off the counter. *Out* and *off*, the two appropriate words in these requests, are quite similar in form to *clean*. All three involve two extended flat hands in active contact with each other.

Finally, some of Koko's overgeneralizations might be attributed to laziness. Even though she knew a number of names for different animals, Koko went through a period when she would name any animal she came across as a *bird*.

In Koko's overgeneralizations, there is evidence that as she acquired words, she was actively involved in investigating the rules that governed her new tool. It was not simply that when she learned the word *corn*, she learned the specific label for a specific object. Rather, she saw that the word corn described small, edible kernels. And so, quite naturally, a couple of months later when she saw an open string bean and some peas, she labeled each as corn. A few months later we gave her some pomegranate seeds, and these too she called corn. What we can see here is Koko demonstrating a capacity for analysis that she put to more sophisticated uses as she grew older. For instance, asked to describe a cut pomegranate, Koko called it *red corn drink*.

Similarly, Koko's overextensions that related to the form of a gesture give the impression that she was exploring the relationship of the motions and configuration of a sign to the thing it represents. They suggest that Koko was trying to understand the structure of the system she was using. Here, too, Koko has become more sophisticated as she has matured.

Koko's overextensions show that she has not been the passive subject of Project Koko, but rather has some predisposition to think and analyze. The ways in which she learns about the language are similar to the ways human infants experiment with language, particularly deaf children learning the same language Koko is acquiring. Overextensions and other language experimentations reveal the interplay between production and comprehension that goes on as the infant explores its newly learned words. In the course of language development the ability both to produce and to understand words changes. Just as there is baby talk before there is full-blown speech, there is baby-think before fully elaborated adult thought. If it was clear Koko was *producing* words, it was still important for me to find out how well Koko *understood* the words she was using, and whether her understanding of language developed similarly to that of children.

11

Comprehension: The Basics

A mynah bird owned by a gas station attendant in Los Angeles is reported to greet visitors with the memorable words, "Up yours, you jive turkey." While it may enjoy the effect its greeting has on the startled humans, it probably does not realize the literal meaning of its message. Language, by all definitions, involves not just the ability to produce words that mean something to the listener, but also the ability to comprehend what one is saying and what one hears. Achieving a desired effect through an intentional message does not mean that one has mastered language. As one critic of the language experiments with apes, John Limber, put it, "Getting someone to bring you a drink is surely not equivalent to telling someone that you want a drink." Say that an infant learns this progression of utterances: —"ju," "juice," "want juice," "want drink juice," "me want drink juice." All will probably produce what the baby wants, and all *mean* the same thing but do not have equivalent grammatical structure. At some point in the progression from "ju" to "me want drink juice," a child can be said to be using language. But linguists disagree on exactly where that point is.

How well does Koko understand language? The question of comprehension is a delicate one because it is risky to assume that the listener—gorilla or human—understands what you are saying even if the responses are appropriate. This question is particularly tricky when the listener is a baby, because, as developmental psychologists and neurolinguists have discovered, the infant's thought, whether gorilla or human, progresses through different stages at different points during its development. At each stage the infant becomes able to use abilities that are only implicit in its earlier actions.

At birth a child's brain is only 40 percent developed. The human brain does not take its final physical shape until about age two. Even then certain interconnections are not completed until as late as seven. What this means, according to some neurologists, is that the six-month-old infant who babbles is not struggling to form speech sounds: the interconnections that permit the associations which are the basis of speech sounds have not yet been completed. On the other hand, at six months some infants have sufficient control of their hands to form words in the form of gestures.

Even after most of the physical development necessary for speech has been completed, the child's brain still must mature considerably before it can handle a number of higher mental functions. Neurologist Norman Geshwind has suggested that in certain cases learning problems in children that are diagnosed as permanent disabilities really reflect nothing more than an abnormally slow physical maturation of the brain.

A child first begins to gauge its world mentally rather than physically between eighteen months and two years old. Jean Piaget believed that this is the time when an infant's ability to understand symbols and form propositions first appears. As was the case with many of Piaget's major contributions to our understanding of child development, the inspiration for his theories came from observing his own children's dawning propositional abilities.

When his daughter, Lucienne, was sixteen months old, Piaget played a game with her in which he placed a gold chain in an opened matchbox. After letting Lucienne reach in and pull the chain out a few times, Piaget closed the matchbox slightly so that Lucienne could still see the gold chain, but could not reach in to pull it out. Lucienne studied this problem intently, and then opened her mouth, first slightly, and then wider and wider. She was doing with her mouth what she wanted to do with the matchbox. After completing this bit of simulation, Lucienne opened the matchbox without hesitation and retrieved the gold chain.

The point of this was that Lucienne, not yet having symbols (words) through which she might phrase her proposition (that by opening the box she might retrieve the chain), performed the simulation physically. Still, she was forming a proposition. The fact that Lucienne had to generate her proposition physically spotlights the economy of language: it requires less energy to make a gestural word than a physical simulation of an action, and less energy still to say a word. It is within this propositional control of the muscles that the seeds of human intelligence are contained.

Although children at Lucienne's age can create simple propositions, they have yet to acquire most of the concepts that will allow them to state or understand the complexities of a sentence like "When did George give that dog a kick?" Similarly, children will not understand plurals, auxiliary verbs, or temporal words before their minds are ready to accept them. And as they learn these concepts, they will refine them in identifiable stages.

Children also use some types of analysis before they understand them. Until age four children overgeneralize words and concepts (e.g., calling any moving vehicle a car), extracting attributes from different words and gradually refining their meanings through experimentation. Children at this stage do not consciously understand what they are doing. By age six, children develop the notion of arranging objects by abstract attributes. Eventually the child will be able to think in an abstract way about the nature of thought.

Between three and twelve children develop control of and learn about their analytical abilities. But it is unclear when the brain has developed sufficiently to allow the child to discover certain analytical abilities. How children's cognitive abilities develop once their "hardware" is in place is subject to a certain degree of chance. The most sophisticated cognitive abilities, such as understanding higher math, seem to flower through a mysterious chemistry between children and their surroundings.

Language is not a sophisticated cognitive ability. For one thing, all unimpaired humans learn it, and to learn language one doesn't need to be able to explain it. However, as three-year-old children refine the meanings of the words they use, they are also refining their understanding of the rules of the language they are learning. When the language is English, the child gradually learns to use plurals, temporal words, questions, tenses, and relative clauses, in predictable stages. If Koko was to learn language I would expect a similar staged acquisition, and an understanding of those components of language that are essential to sign language. Establishing whether Koko comprehends language, however, is made difficult by the fact that there is no consensus at the moment about when a child can be said to have acquired language. Nor is it clear whether the components of English that a child learns in stages are inherent in all language or are peculiar only to English, or to speech in general.

In the absence of such consensus, the best course in starting out seemed to be to investigate Koko's use of some of the various forms (such as questions) that scholars cite as critical for language, to see whether she was capable of demonstrating them and understanding their significance. Then we would leave it to others to determine what was or wasn't language.

In my project, I could not isolate Koko from spoken English, given the circumstances of her infancy. Moreover, I realized that if she did learn to understand English, that understanding would serve as an excellent corroboration of her understanding of words in sign language.

At the beginning of the project, there was little evidence to suggest that a gorilla could learn to understand English, and several scientists believed that if an animal could not generate spoken words, it did not have the necessary equipment to understand them either. This is a rough description of the motor theory of speech perception.

If Koko learned to understand English it would not only help us to establish Koko's intent, but would also comment on some of the assumptions about the nature of comprehension itself. And it would disprove the motor theory of speech perception and the corollary argument that it is the child's innate capacity for auditory analysis that distinguishes him from the great ape.

Still, the principal benefits of learning English would be to Project Koko. If Koko could translate from English to sign language it would prove that she understood the symbolic nature of language. For instance, if Koko made a gestural rhyme on a word in sign language, we could then see whether she understood the concept of rhyming by asking her to sign a word that sounds like another spoken word. If Koko showed that without prompting she could associate words by gestural similarity, that would show that her sign homonyms were not merely mistakes, but rather evidence of a sophisticated understanding of the underlying structure of language.

Of course, at the beginning of Project Koko, possibilities such as this did not enter my mind. I had no clear conviction that Koko could learn to understand English. Moreover, despite the fact that English was used along with sign language, Koko's language instruction was exclusively in sign. If she was going to learn English, she had to pick it up herself. To do this, Koko would have to separate her understanding of the meaning of the word from its expression in sign, and then use this understanding to establish relationships between the gestures she was taught and the sounds she heard. This would be expecting a lot from a species that had never before been taught any language.

Koko's exposure to English actually predated her first lessons in sign language. During her infant illness she had spent some weeks in the homes of the Reuthers and the Bells, and at the time I began to work with her she was already responsive to "no," "Koko," and several other spoken words. This was encouraging. By the age of two-and-a-half, Koko would occasionally act out things she heard on tape. Once I was transcribing an audio tape in her presence and was surprised to see her starting to break a spoon just after the words "broken spoon" sounded from the tape. Soon after that, she startled us by spontaneously signing the word *candy* when she heard a visitor speak it. From then on she would regularly translate English words and phrases that she heard. She also began to answer questions posed to me. One day a visitor asked what the sign for *good* was. Before I could respond, I noticed Koko making the sign *good*.

After a few instances of such gorilla eavesdropping, we began to be more careful about what we would say in Koko's presence. I once made the mistake of telling one of her teachers that Koko had done particularly well during a lesson. Koko began charging about, displaying and misbehaving. Then I revised the praise, and said Koko was stupid. Koko laughed and calmed down.

We have come to use the time-honored device of spelling rather than saying key words when we don't want Koko to know what we are talking about. Somehow Koko has figured out that "c-a-n-d-y" spells one of her favorite treats, so that now we have to use even more artful subterfuges when discussing such highly charged topics.

Koko's growing understanding of English had its practical uses as well. If my hands were otherwise occupied, I could still tell Koko to clean up her room and have her respond correctly by fetching a sponge and wiping up a mess she had made.

The obvious question raised by Koko's appropriate responses to English was whether she in fact understood the spoken messages or instead made good guesses by interpreting nonlinguistic cues such as body position, tone of voice, or the direction of my gaze. When she was four-and-a-half, I sought to determine Koko's relative comprehension of English, sign language, and simultaneous communication by administering a test called "Assessment of Children's Language Comprehension," or ACLC. It is important to note that on neither this test nor any other test of comprehension did Koko receive any drilling or training. The test consists of forty cards printed with line drawings or silhouettes describing different objects, attributes, and relationships between objects (see Figure 3). The first ten cards depicted vocabulary items, and the remaining thirty cards tested Koko's comprehension of phrases of varying complexity. The phrases had between two and four critical elements—e.g., "point to the bird above the house," or "point to the broken sailboat on the table." In the vocabulary item test, Koko would have to point to the appropriate item; in the comprehension section, she would have to choose which scene accurately depicted a statement.

We gave Koko the test under three different conditions—sign only, voice only, and simultaneous sign and voice. To ensure against cueing, we administered the sign-only and voice-only tests blind and videotaped Koko's responses. This meant that the person who administered these sections of the test could not see what Koko was responding to. After Koko had made her choice, the experimenter (I or one of my assistants) would look

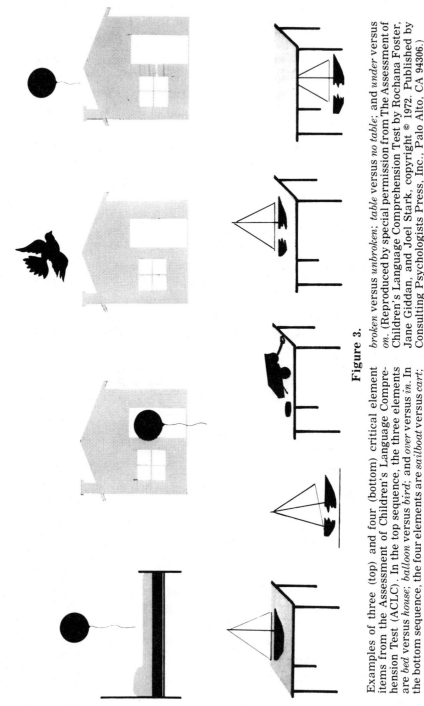

Figure 3.

Examples of three (top) and four (bottom) critical element items from the Assessment of Children's Language Comprehension Test (ACLC). In the top sequence, the three elements are *bed* versus *house*; *balloon* versus *bird*; and *over* versus *in*. In the bottom sequence, the four elements are *sailboat* versus *cart*; *broken* versus *unbroken*; *table* versus *no table*; and *under* versus *on*. (Reproduced by special permission from The Assessment of Children's Language Comprehension Test by Rochana Foster, Jane Giddan, and Joel Stark, copyright © 1972. Published by Consulting Psychologists Press, Inc., Palo Alto, CA 94306.)

at the card and see whether Koko had been correct. If she was correct, Koko was given a little treat.

The results of Koko's performance on this test were encouraging (see Tables 2 and 3). She performed slightly less accurately than educationally handicapped children. She performed as well in the sign-only tests as in voice-only, even as the problems became harder. This indicates that something other than conceptual difficulty was influencing lapses in her performance, since if that were the case her performance would deteriorate as problems became more difficult. From her response to other tests, it would be fair to infer that her errors stem partly from the gorilla's hatred of regimented, boring tests. Even so, Koko performed significantly better than chance, and, as one might expect, she responded slightly better to simultaneous communication than to either sign or voice alone. What was surprising was that she responded equally well to English without sign as she did to sign. At the most difficult level she performed about as well as children of her age who were either neurologically or educationally handicapped. However, her performance was distinctly inferior to that of a normal hearing child.

Besides administering the ACLC, I tried other procedures to see how Koko's comprehension and statements were affected by the language mode her companions used. Over eleven months in 1974, I sampled Koko's statements in circumstances in which she was exposed either to sign language only or to spoken language only. I also kept tabs on how my statements were affected by which language I was using. One thing I discovered was that I was considerably more loquacious when

Table 2. Koko's Performance on the ACLC

| Number of Critical Elements | Percent Correct | | | | |
	Chance	Sign and Voice	Sign	Voice	Total
One	20	72			
Two	25	70	50	50	56.7
Three	25	50	30	50	43.3
Four	20	50	50	30	43.3
Totals (two, three, and four elements)	23.3	56.7	43.3	43.3	47.8

Table 3. Performance on the ACLC by Normal Children,
Children Diagnosed as Neurologically or Educationally
Handicapped, and Koko

Number of Critical Elements		Percent Correct			
	Chance	Normal Children	Handicapped Children	Koko (sign and voice)	Koko's Average (of sign, voice, and sign and voice)
One	20	91.1	91.1	72.0	
Two	25	91.1	87.5	70.0	56.7
Three	25	83.7	74.4	50.0	43.3
Four	20	64.1	43.8	50.0	43.3

using voice than when using sign. On the other hand, Koko's performance was unaffected by the language I used. Her sign utterances occurred just as frequently and with no difference in length whether I was using speech exclusively or sign exclusively.

Another indication of Koko's ability to process English is her translation of spoken statements into sign. While simple imitation of a gesture might only indicate mimicry of a movement without a conceptual grasp of its meaning, imitation through translation of a spoken word to gesture would indicate that Koko comprehends the common meaning expressed through the two different languages.

Her imitations of both speech and sign fell into four categories. If I said or signed *big glass milk*, she would either repeat a word or phrase completely (*big glass milk*); or reduce it to its telegraphic essentials (*big milk*); or expand a phrase (*big glass milk drink*); or paraphrase a statement (*glass milk drink*). As in the case of her responses to questions in different languages, Koko's translative imitations of voice were about equal in number to imitations of sign, and slightly higher when simultaneous communication was used. The samples in question were taken between ages two years eight months and three years six months. Over these eleven months Koko did show an increasing tendency to copy statements made by others. In all she copied roughly 10 percent of the statements made by the experimenter, but toward the end of the eleven-month period Koko was imitating more than twice as much as at the outset. This was also the period during which Koko's vocabulary began to show the greatest growth, and we might see a connection between this increased imitation and the phenomenon called learning to learn, in which the infants begin to understand the larger dimensions of what they are being taught.

Curiously, Koko's utterances were longer when imitating statements made in voice only. Perhaps the mental exercise of translating from auditory to sign also stimulated Koko to elaborate on the content of these imitated utterances.

How accurate were Koko's responses to the normal questions, requests, and instructions humans put to her? In answering this question, I analyzed Koko's responses, separating them into six categories: verbal correct (meaning correct signs); verbal incorrect; nonverbal correct (correct actions); nonverbal incorrect; termination (Koko turns her back); and no response (Koko continues as before, without giving any indication whether she has noticed that something has been said to her).

The results confirmed the suggestion of the ACLC test and Koko's imitations that she understood to some degree the languages being used. Again as in the other instances, Koko responded slightly more frequently and accurately when simultaneous communication was used than with either sign only or voice only. Koko's responses were correct about 58 percent of the time when addressed exclusively in sign or voice, and 64 percent of the time when simultaneous communication was used. This suggests that Koko understands English at least as well as she understands sign language. It is important to keep in mind that she has had no formal instruction in English. Her English comprehension was likely facilitated through the transferred application of her understanding of the abstract properties of sign language.

The fact that Koko translates words and phrases from English to sign language and responds appropriately in sign to her companions' spoken English suggests that Koko understands the language used in her presence. But it gives no clear idea of how many of the conventions of the language, such as question forms, she understands.

An important event in a child's acquisition of language is the gradual comprehension of interrogative words such as "who," "what," "where," "when," "why," and "how." The mastery of the question has a vast utility for children, as any parent knows who has been driven crazy by the persistent interrogations of a four-year-old. Questions allow children to actively guide their acquisition of language and knowledge. The mastery of question forms indicates that children have begun consciously to grasp the utility of language. They have undergone what Piaget called a miniature "Copernican revolution" in which they discover that they are not the center of the universe. They have begun objectively to study the world that they now know is separate from themselves.

Susan Fisher, who has compared the development of language in hearing and deaf children, believes that both groups acquire the interrogative "wh" forms in stages. She borrows her idea for the staged development of question structure from psycholinguist Roger Brown, who first put forth the idea that children acquire language in several stages. For hearing children, Stage I involves posing questions through intonation with only sparse use of "wh" words. In Stage II, children will begin to use question words, even before they learn auxiliary verbs: "Where my glove?" or "What book name?" Finally, in Stage III, children learn to alter sentence structure, inverting subjects and verbs to pose questions—"It is my book?" becomes "Is it my book?"

According to Susan Fisher, the development of questions in deaf children is somewhat different, although also in stages. Because sign language permits the child to give a sentence an interrogative cast merely by holding a sign or by accompanying a statement with a questioning look, deaf children master this form of questioning in the first stages of language acquisition. However, deaf children master the use of "wh" questions much more slowly and use them much less frequently than do hearing children. Because there are simple ways to turn a gestural statement into a question, posing a "wh" word often becomes redundant. Thus in the second stage deaf children learn "where" (which has obvious utility), but use the gestures for "who" and "what" only in limited ways. Like hearing children, who find combining a "wh" word with a subject-verb inversion —"Why is it?"—too difficult, deaf children at this stage tend to drop the "wh" word when forming a complex sentence. By the third stage the deaf child responds appropriately to "wh" questions with the "wh" word at the beginning of the sentence.

Ultimately, the deaf child learns to use "wh" words, although again it must be stressed that the deaf do not use question words as often as those who speak.

Unlike children, Koko does not use "wh" words except on rare occasions. There are several possible explanations for this. Koko does frequently ask questions through acceptable gestural intonation or inflections. For Koko, the serviceability of this form of questioning is enhanced because her human companions are perhaps unusually attentive and willing to respond to her slightest interrogative movements. Others might argue that Koko's infrequent use of "wh" words indicates that she does not understand them. However, this argument is undermined by the fact that she responds to them appropriately.* The one "wh" word Koko has most difficulty with is "when", a word that poses problems for all children.

Comparative data on child and chimp understanding of questions show that children at Stage III—between twenty-two months and forty-two months old—answered "who," "what" and "where" questions significantly less well than Washoe did at age five.

With Washoe's and the average child's performance in mind, I sought to determine whether Koko's comprehension would be comparable. One major difference between my study and those on Washoe and children was that, as in my other studies of comprehension, I collected data in English comprehension and sign comprehension as well as in simultaneous communication. I also included questions—what happened, where + action, and why + action—that were not in the Gardners' study. Table 4 shows the question types, sample questions, and examples of Koko's replies.

With regard to Koko's performance, so far I have only been able to analyze a small part of the massive amount of data collected. I have done an intensive analysis on data taken from one month in the summer of 1977, when Koko was six years old. During this period Koko was asked randomly selected questions from the fourteen question types. The mode—sign, voice, or simultaneous—was also varied randomly. The total number of questions asked during the month was 427, which made the

*It is possible that Koko's paucity of "wh" questions is an individual idiosyncrasy. Our other gorilla, Michael, does frequently use the word "what" when asking questions.

Table 4. Relation Between Questions and Answers

Question Type	Sample Question	Koko's Reply
Who + pronoun	Who you?	Koko.
	Who me?	Think Barbara.
Who + action	Who eat?	Ann.
	Who see?	Know Mike there.
Who + trait	Who smart?	Smart Koko.
	Who funny?	Penny.
Whose + demonstrative	Whose bug?	Bug Koko.
	Whose this berry?	For Koko red berry.
What color	What color this?	That red.
	What color this?	Koko think orange.
What + demonstrative	What this?	Think dog.
	What that?	Meat.
What do now	What now?	Make sandwich.
What want	What want?	Grape gimme.
What happened	What happened?	Me break.
Where + action	Where bite?	Bite ear.
	Where tickle?	Underarm.
Where + object	Where bird?	There bird bird.
	Where berry?	Do berry under.
How + state	How do you feel?	Fine.
	How does it taste?	Good.
Why + action	Why pinch Cindy?	Food hurry.
Why + state	Why sorry?	Bad me.

question sample size roughly comparable to the sample used in Project Washoe.

Koko's vocabulary in answering the questions drew from a set of categories roughly similar to Washoe's. She used nouns, proper nouns, modifiers, markers (temporal words, negatives, imperatives, expletives, demonstratives, prepositions, and so on), verbs, and locatives. For the purposes of comparison, I used the same criteria for analyzing Koko's response as the Gardners did. That is, only if Koko used a sign from the specific target category for the question would Koko's answer be accepted as correct. I further categorized her answers as appropriate or inappropriate depending on the context of the question. Thus Koko might give an answer that was grammatically incorrect but appropriate, or correct but inappropriate (such as responding *red* when the question was *What color?* and she was shown a blue napkin). In other words, her answer might hit the grammatical target specified but be inappropriate because the answer was untrue.

Like Washoe, Koko performed better than children at Stage III in answering questions. Her answers were grammatically correct 83 percent of the time. In those question categories which Washoe answered correctly 84 percent of the time, Koko's accuracy was 88 percent. It turns out that Koko more often hit the grammatical target category correctly than the appropriate (accurate) answer, which surprised me because Koko's teachers tended to reinforce accuracy rather than grammaticality. Koko got both right about 70 percent of the time. She seemed to falter in answering questions involving "What happened?" or why + action. Perhaps this lapse occurred partly because these questions often involved scolding Koko for some misadventure—"Why did you do that?"—and in these circumstances Koko often tried to change the subject or leave the scene of the crime.

I then analyzed Koko's performance in four major question categories: where, who, what, and why/how. Koko's performance was poorest when explaining the "why" of something—the most nebulous and analytical category. She also responded less frequently to "what" questions. These questions, when not involving scolding, often called for a recitation of vocabulary items, which frequently made Koko impatient.

Koko's understanding of English raised the question of whether it was possible to construct a device that would permit her to speak English as well. Patrick Suppes, a logician and head of Stanford's Institute for Mathematical Studies in the Social Sciences, and his colleagues designed a keyboard-computer linkage that enables Koko to talk by pressing buttons linked to a voice synthesizer. The buttons are on a sturdy computer console. Each of the 46 buttons represents a word and is painted with arbitrary geometric shapes in one of ten randomly chosen colors. When Koko presses a button, an unintoned female voice utters the appropriate word through a speaker. Simultaneously, the word is recorded in a computer data file, so that we can later analyze Koko's utterances for consistencies of usage and word order.

Koko quickly learned to use the keyboard. She became quite adept at typing out requests such as *Want apple eat* and *Want drink sip.* A one-finger, hunt-and-peck typist, Koko usually keeps one hand free to sign statements as they are uttered through the voice synthesizer. Koko tends to repeat many of the words as she taps the keyboard, producing statements such as *Apple hurry apple apple go go go go eat eat eat Koko Koko Koko.* The synthesizer has frequently malfunctioned, and although we have collected an enormous amount of data, we have not yet had time to analyze Koko's "spoken" language in detail.

The purpose of the various comprehension tasks I set for Koko was to develop some bedrock data on her understanding of the words she was using. The implication of this data—that Koko understands what a symbol is—has been corroborated by the more sophisticated comprehension tasks we set for Koko as the project developed. These (which will be discussed in Part III) involved free association, metaphor, rhyming, inventing signs, and so on. Such abilities suggest a far more elaborate understanding of sign language than do the basic comprehension tests discussed in this chapter, but without this basic data, I could not have any confidence that Koko was in fact doing what she seemed to be doing in more sophisticated uses of Ameslan.

Apart from the question of whether Koko understands the symbolic nature of the word, there is another nettlesome problem concerning Koko's understanding and use of sign. Is Koko merely randomly generating words that relate to different aspects of what she is trying to describe, or does she have the capacity to deal with grammatical conventions—evident in word order in particular—that allow us to link symbols in meaningful and infinitely creative ways?

12

The Troubling Question of Word Order

Since the Gardners first published their findings on Washoe, much of the debate about ape language use has focused on the issue of word order. According to some linguists, the grammatical conventions manifest in word order in English are what allow a language to be open, productive, or creative. Through the syntactical devices expressed in word order, one can generate a limitless number of different but still meaningful sentences. With some naïveté, critics assumed that word order had the same significance in sign language that it did in English. Some of these criticisms were eventually recanted or modified. But the question of whether or not an ape is capable of demonstrating the grammatical forms that in English take expression through word order remains unsettled.

To talk about an event displaced in time and place from a given discussion (e.g., "last week I saw an elephant at the circus"), one needs symbols to represent the objects and attributes being discussed. One also needs a framework on which to hang those symbols so that the relationships between them may be meaningfully re-created. In English the set of conventions that govern word order provides such a framework. Thus, for many critics of the language experiments with apes, the ability to name is not sufficient to justify the claim that an animal has language. Rather, the animal must show that it understands the grammatical linkages that allow one to combine words in varied but still meaningful ways. Determining whether Koko understood these grammatical linkages would seem to be a simple matter of testing her comprehension of sentences of increasing complexity, as well as analyzing her utterances to see whether she was combining her words in a rule-governed way. A number of difficulties, however—some having to do with the differences between sign and spoken language, and others having to do with questions about the significance of word order—muddy the waters on this question.

No animal or human could be said to understand the nature of *English* unless the creature also understood the significance of word order. I emphasize the word *English* because, as noted before, word order has different significance in different languages. In English, word order is the basis of structure, and, ultimately, a proof of a speaker's intent. It is *a* proof rather than *the* proof because, as we all know, not everybody speaks grammatically all the time. In many cases we interpret the meaning of a speaker's remarks according to a variety of nonlinguistic cues such as expression, tone of voice, speech rhythms, and gesticulations. Thus, we often know the *sense* of what someone is saying before he finishes a sentence.

Often people don't finish sentences. And, as countless therapists and self-improvement guides have stressed, when people are speaking, a lot of information is passed on that has very little to do with the literal meaning of what is being said. The most innocent message can be charged with the darkest and most threatening meanings. Sentences that are virtually indecipherable in transcription can carry powerful "unspoken messages."

All of these qualifications should put the matter of word order into perspective. Linguists who isolate grammar (or "deep structure") as the critical element of language are guilty to some degree of looking at language through the wrong end of the telescope. In emphasizing the importance of grammar over all other aspects of language, these "deep syntacticians" have tended to lose sight of the other factors that condition human communication. However, to mention these qualifications is not to minimize the importance of word order in English.

Say you want to describe to a friend something that happened to a distant acquaintance you have in common: "I read that that fellow who tried to telephone your grandfather got hit by a train last week." Reading or hearing this, you realize that it was not the I or your grandfather but the fellow who tried to telephone your grandfather who was hit by a train. We know that the phrase *I read* conditions the remainder of the sentence, that the first *that* conditions the following relative clause, and that the phrase *tried to telephone* only conditions the fellow mentioned in the relative clause. The conventions of English—flagged in this sentence by the words *that* and *who*—indicate to the listener what parts of the sentence the various verbs act on. The word order might be substantially altered in this sentence and still yield the same meaning: "I read that last week a train hit that fellow who tried to telephone your grandfather."

If one gave equal weight to each word in the above sentence it would be meaningless. What word order and the other conventions of English signal is an ability to recursively follow rules—that is, in a patterned, recurring way. This means that the person hearing a sentence will use a set of rules to understand a noun phrase constituent of a sentence like the one above and then apply a set of rules to the noun phrase as a whole in a similar operation performed on the entire sentence. This recursive ability is quite possibly the most pervasive and powerful attribute of human thought. It basically allows us to mentally take apart the world and put it back together again. Noam Chomsky, the leading proponent of deep structure, has said in lectures that he has not seen evidence that apes possess this recursive rule-following behavior. In fact, Chomsky has stated that he sees no more evidence of language in ape use of symbol systems than he sees in a child waving its arms as evidence of incipient flight. It should be mentioned, however, that there is no consensus among linguists about the universality of deep structure.

These aspects of what is called "sentence theory" have become the rallying point for those looking to isolate what is unique in human language. Moreover, because apes have demonstrated recursive rule-following behavior in nonlinguistic areas involving tests of pure reasoning ability, an old hypothesis of Noam Chomsky's has enjoyed a revival. This idea is that man is genetically equipped with a set of syntactic instructions which permit the generation of all languages and which are separate from the structures permitting other higher intellectual functions. Besides the ape's presumed failure to learn syntactic structures like word order, proponents of this hypothesis also cite the speed and regularity with which children learn language despite vast differences in intellectual capability and cultural background.

This hypothesis contrasts sharply with Piaget's ideas discussed earlier. In Piaget's framework, language develops out of general cognitive structures. The child's sense of syntax develops as the child inductively refines rules; a special set of genetic structures permitting syntax is not needed. As we will discuss, the ways in which Koko introduces complexities into her utterances suggest that there are alternative perspectives to Chomsky's on the question of the ape's syntactic capabilities.

In Project Washoe, the Gardners were not looking primarily at Washoe's word order, although their subsequent analysis of Washoe's combinations showed that as time went on, Washoe tended to adopt the English word order used in sign by her human teachers. Roger Fouts discovered that one of his chimp pupils, Lucy, understood the difference between sentences like "Roger tickle Lucy" and "Lucy tickle Roger." Another of Fouts's studies demonstrated that a chimp named Ally understood the significance of word order in prepositional phrases. David Premack's chimp Sarah seemed to respond appropriately to complex sentences, but it is possible she was cued inadvertently—and responding to sentences is quite different from generating them.

A manifesto for a new generation of critics of ape language research has been Herb Terrace's revisionist critique of his own experiment with a chimp named Nim. Terrace sifted through some 19,000 two-word combinations generated by his chimp pupil Nim (full name Neam Chimpsky). What came out of this analysis was a critique that would lead one to believe that neither Nim nor any other ape has even a rudimentary understanding of the nature of language.

First Terrace performed a distributional analysis to see whether Nim's combinations were random or showed regularities in sign order. When this analysis showed that Nim did indeed seem to have strict preferences as to the order in which he would combine signs, Terrace then applied a series of "filters"—different types of analysis—to his data to try and determine what strategies might be influencing Nim's choice of sign order.

After eliminating randomness and rote as possible explanations for Nim's choice of sign order, Terrace at first felt that there was some basis for believing that Nim was choosing his word order with the intent of establishing syntactic relationships between his words. However, later, after Nim had been returned to a primate facility in Oklahoma, Terrace and his colleagues studied the three and a half hours of videotapes of Nim's utterances and decided that it was premature to claim that Nim understood the basis of sentence structure.

One thing that troubled Terrace was that there was no real increase in the length of Nim's utterances as the chimp learned more and more signs and more about the language. "Apparently," wrote Terrace, "utterances whose average length was 1.5 signs were long enough to express the meanings that Nim wanted to communicate." Moreover, Terrace noticed that even in Nim's longer, three-, four-, and five-sign utterances, the additional signs did not serve to express complexities not expressible in two-sign utterances, but rather served merely to add emphasis to a statement adequately expressed in a two-word combination.

Two other things caused Terrace to question Nim's linguistic competence. One was the small number of signs he would use in expressing various semantic relationships. Another was Nim's large number of imitations. When reviewing videotapes, Terrace noticed that Nim often imitated what his teachers were saying to him. Conversely, fewer of Nim's utterances seemed to be spontaneous than Terrace at first believed. Only 10 percent of Nim's videotaped remarks were spontaneous. Nor does Terrace feel that Washoe or Koko are any more spontaneous than Nim. On the basis of a film of Washoe done for the *Nova* television science series, in which Koko as a baby appeared for fifty seconds, Terrace claims that Washoe and Koko received similar prompting. Terrace cannot justify such conclusions on Koko and Washoe on the basis of brief snippets of film from the *Nova* series. Any commercial film involves noise, light, and bustling camera crews that have their effects on an ape's—or for that matter, a human's—state of mind.

Although Terrace argues that Nim has not generated true sentences, methodological problems with the experiment and questions about Terrace's analysis of his own data make it difficult to draw conclusions from his study. Terrace's critique of Nim's combinations leaves the impression that the chimp is unspontaneous, uncreative, and unstructured in most of its utterances. In Koko's case, however, there is reason to believe that a great deal of creativity, spontaneity, and structure characterize her utterances (although creative elements that contribute additional meaning, such as modulation, would be missed by the "filters" Terrace has used on Nim's combinations). For instance, Terrace reported that 13 percent of Nim's utterances were spontaneous; samples of Koko's statements show that an average of 41 percent are spontaneous. Terrace found 39 percent of Nim's responses were complete or partial imitations of his teachers; only 11 percent of Koko's utterances are imitative in this way. And, of those imitations, only 7.3 percent of Nim's were expansions of the teachers' statements, while 36 percent of Koko's are expansions on her teachers' statements.

Koko's vocabulary is significantly larger than Nim's, and the mean length of her utterances is longer—roughly 2.7 compared to Nim's 1.5 signs per statement. This is still a low figure compared to children's sentences, and one that makes it difficult to examine word order. But although analyses performed to date indicate that the ape adds relatively little complexity to its statements through additional signs and identifiable word order, as speakers of English do, that does not mean the ape is incapable of increasing complexity. In sign language there are ways of adding complexity to a sentence without adding length.

Which conventions in English are basic to all languages and the meanings they express? Which conventions in English are related to the constraints of speech? Have theories confused devices for processing sound information with universal properties of language? These questions have been only marginally considered, even by the most distinguished linguists, yet they rise to prominence when we begin to contrast spoken language with sign language.

A basic difference between spoken language and sign language is that while it is possible to add meanings to a spoken word through intonation, in most cases to amend a meaning or express complexity one must generate a number of additional words in an appropriate sequence. Spoken language is therefore linear and sequential. In contrast, it is much easier in sign language to present a whole thought simultaneously. The importance of this difference should not be underestimated. It means that a spoken language requires a number of structures that signal the relationship of the words yet to be spoken to those which have already been uttered. Because of the sequential nature of a voice message, the individual must also be equipped with a highly developed attention span—and one of the most dramatic contrasts between the young ape and the young human is attention span. Sign scholar Louis Fant has observed that in short sentences of 3 to 4 signs, there are few constraints on word order (3 and 4 being the outer range of the majority of ape utterances—although Koko frequently makes much longer statements). But in longer sentences, signs are arranged according to time sequence of events. For example, "It was a thrill to watch the sunrise this morning" would be translated *Now morning, sunrise, I look-at, thrill.*

Although Ameslan lacks inflections, by modulating where or how a sign is made, the deaf express grammatical functions simultaneously that are expressed by sequential devices such as word order in spoken language. Koko sometimes varies the motion of a sign to indicate a specific actor. When she moves the *sip* sign away from her mouth toward me, Koko is actually saying "You sip." Or Koko will create a compound sign by simultaneously modulating two signs. "I love Coke," for instance, she says by placing her arms in the huglike *love* position as she makes the *Coke* sign with one or two hands.

Typically, Koko will selectively modify one of the formational components—the configuration of the hand, motion of the sign, or place where the sign is made—while holding the others to standard form. This is also characteristic of modulations humans perform on standard signs. It suggests that Koko has a grasp of the underlying structure from which signs are generated, and it indicates that Koko uses Ameslan in the "open" and "creative" way that Terrace thought was critical if an ape was to be credited with language.

I have not yet done a full analysis of Koko's word order. A preliminary study I did as part of my doctoral thesis revealed that there are consistencies in the way Koko combines words. During one twelve-month period in 1974, Koko placed the modifier in the initial position of her two-word statements 75 percent of the time. On the other hand, she would consistently place a special type of modifier, the demonstrative (*this, that*), in the second position. There were similar consistencies in other categories of her statements.

Over time certain patterns have changed. While Koko at first placed adjectives before nouns (*red berry*) as we do in English, by age three-and-a-half this pattern reversed, and has since remained consistent. Utterances such as *lipstick red* and *baby new* are now the rule, as is the case in Ameslan, even though most of her companions model the English order. The difficulty is determining whether these consistencies are derived from some set of rules Koko uses to arrange her utterances, or whether contextual, physical, or other factors affect her sign order. I look forward to the day when I can do a thorough study of the hundreds of thousands of statements we have collected from Koko throughout the project.

When critics write of the patterned rule-following aspects of language, they tread in a poorly defined area where language and other aspects of propositional thought merge. There is a logic to language conventions that permits us to utter a meaningful sentence—and it is independent of the logic in whatever it is we are saying. One can, of course, utter nonsense that is grammatically flawless, such as "The barefoot boy with shoes on stood sitting on the chair" or "Colorless ideas sleep furiously."

However poorly understood at this point, sign language may offer some insights into the relationship between the various paths in which man's propositional abilities developed. In part because of the language experiments with apes, the old hypothesis that before man used articulate speech he communicated in a language of gestures has been revived. This hypothesis offers a means of linking the development of language with the flowering of man's other propositional abilities.

13

Sign Language

When most people think of the language of the deaf, they think of finger-spelling, but finger-spelling is not a language, it is merely a method of translating a spoken language into gestures. Ameslan is a language in itself. It is not a mapping of English. There are obvious constraints operating on sign language that do not apply to spoken language. For instance, people of equal intelligence would have a much smaller vocabulary in sign language than in spoken language. Few sign-language dictionaries have more than 2,000 entries, while a standard-sized abridged English-language dictionary might have more than 200,000 entries. What permits the English lexicon to be so large is that it is encoded in writing. Since sign language lacks many of the technical words of English, a more useful comparison would be with the much smaller lexicon of a spoken language that had no tradition of writing. It is also interesting to note that a Stanford student has calculated the average number of different words used in scientific papers at the graduate level to be around 300.

One extraordinary charge leveled at those attempting to teach sign language to apes has been that Ameslan is not a language. Besides leaving us with the implication that the deaf are not human (because if language is the distinguishing characteristic of humans, and Ameslan is not a language, then the deaf who communicate through Ameslan do not satisfy the criterion for "humanness"), the criticism is based upon a misunderstanding of the nature of Ameslan. One of the characteristics of human language is openness or productivity. This means that the communicative system is structured so that a finite number of individually meaningless units (such as the alphabet) can be combined into infinite numbers of meaningful messages. In linguistics this double substructure to a language—phonemes and morphemes—is called "duality of patterning." In contrast to human language, animal call systems are traditionally described as closed: messages are not constructed from subunits, but rather each call evolves as a separate unit; the animal's vocabulary is only so large as the number of individual signals it has developed, and those signals cannot be recombined to generate new meanings.

The reason some critics believe that Ameslan lacks duality of patterning is that they focus on certain signs that are iconic. That is, some signs seem to sketch the outlines of the object they represent. This has led to the charge that Ameslan is not really a formal language, but rather a ragtag collection of emblems in which a sentence is little more than a pantomime. This criticism has been refuted by a number of sign-language scholars, who point out that most signs are constructed of four distinct elements—location, configuration, position, and movement. One of the more interesting refutations comes from the unexpected quarter of an early critic of Project Washoe: Ursula Bellugi. After Bellugi retracted her initial criticisms of Project Washoe, she set about to study sign language in depth. She has been responsible for a good deal of the research comparing the development of sign language in deaf children with that of spoken language in normal children.

In answer to the criticism that Ameslan is an elaborate pantomime, she uses the example of the sign *egg*. In a study, Bellugi asked twelve people to convey in pantomime certain objects for which there were also signs in Ameslan. For *egg*, people made five recognizably characteristic movements— describing the shape, mimicking the action of breaking an egg, throwing away the shell, and so forth. One of these actions turns out to correspond closely to the sign in Ameslan for egg. In Ameslan one says *egg* by bringing the first two fingers of each hand together and then down and apart. This motion seemingly imitates the breaking of an egg, which several of the twelve subjects used in their pantomime. However, Bellugi points out that the sign for egg is an exact movement, and that some of the egg-breaking movements delivered in pantomime would signify words other than egg if isolated. Moreover, she notes that the deaf learn signs according to hand configuration, the movements involved, and the place on the body where the sign is generated. In this regard, *egg* has more in common with such signs as *name, train,* and *short* than it does with its iconic roots. Bellugi also points out that, over time, many signs that have iconic roots develop to the point where their iconicity is unrecognizable. *Sweetheart,* which in 1913 looked like the tracing of a heart on one's chest, with fists together and raised thumbs wiggling, now is just the movement of the two thumbs with the hands placed in front of the center of the body.

Finally, on this issue, Bellugi studied the types of errors made by the deaf in their written responses to signs made in Ameslan. She and her colleagues discovered that just as hearing people would make errors that sounded like the correct word ("horse" for "house"), the deaf subjects would err by mistaking a sign for another that was similar in form, though totally different in meaning. It was clear that the deaf were not translating sign into English and then back into sign, because their errors differed markedly from those of hearing people. Nor were their errors misinterpreted pantomime, since they often meant something completely different from what a pantomime would suggest. Rather, the deaf to some degree seemed to be encoding words according to form rather than iconicity. Nor were they encoding the form as a whole: Bellugi discovered that errors often involved a particular constituent of the form, such as the place on the body where the sign was to be made. This has also turned out to be the case with a substantial number of errors made by the chimps and by Koko.

What Bellugi and others have made clear is that Ameslan is in fact a language, and not in any sense dependent upon spoken language.

At the beginning of Project Koko, much of this work had not been done. It is somewhat ironic and sad that a good deal of the current interest in the nature of sign language stems not from interest in the deaf but from curiosity about the nature of a language that might be acquired by another animal.

The attempts to impart sign language to chimps and to Koko have also enlivened another line of research. The fact that our closest primate relatives can learn a gestural language even though they cannot talk has renewed interest in an old idea— namely, that mankind's first language was gestural, and that speech evolved after man first developed the ability to control his actions purposefully in the form of making tools and using a rudimentary sign language. This hypothesis is compelling because, if true, it suggests a single origin for the disparate collection of abilities common to logic, language, and technology.

The performance of the great apes in sign language, given the physical limitations that prohibit their acquisition of spoken language, is an important piece of evidence in support of this theory. But there is other evidence as well. Deaf parents report that their children generate their first sign at about six months, which is between three and six months earlier than the speaking baby says its first word. As noted earlier, the brain's interconnections necessary for propositional control of the hands are in place earlier than those necessary for speech. Perhaps this early flowering of sign abilities recapitulates an evolutionary stage during which man used a gestural language.

Perhaps the most tempting aspect of the gestural theory of language origins is that it offers an elegantly simple explanation for the origin of both technology and language—and even a teasing glimpse of the possible origins of a sort of "deep structure" in language. Anthropologist Gordon Hewes, a leading proponent of the gestural theory, points to the methods early man used to make tools. About 80,000 years ago in France, man fashioned blade-like Levallois tools from pieces of stone. The construction of these tools was not the mere refinement of a found object, but rather involved a programmed series of actions, hierarchically organized. In other words, there was a syntax or grammar to this series of actions; and, like a sentence full of relative clauses, the final product was not implicit in any of the steps leading up to it. Thus we might envision a common origin for both technology and language, and, indeed, all man's higher mental capacities, in an ordered ability to program motor actions.* For that matter, there are forms of apraxia (damage to brain centers controlling motor actions) that are similar to certain aphasias (damage to the language centers). Hewes has made the point that perhaps the ability to program motor actions is the "deep structure" Noam Chomsky should have been looking for. The whole flowering of the human intellect may flow from some original ability to form propositions through our own hands.

Apart from the attractiveness of the gestural theory as an explanation of the origins of man's peculiar abilities, it also helps to explain the presence of sign abilities in the great apes, and it offers us a common ground for entering the mental world of Koko. If the structure of our mental world works back ultimately to the logic of motor patterns, we might expect that a creature who shows an understanding of the logic of sign language motor patterns would share with us some of the mental states that derive from those abilities. By eliminating those aspects of human thought associated with the flowering of speech, we might someday develop a true yardstick by which to measure Koko's abilities.

In the meantime we have, apart from the literature on psycholinguistics, the totally unsatisfactory yardstick of intelligence tests, most of which, I discovered, measure a jumble of abilities with no sensitivity to the types of distinction Hewes and others are positing for higher mental function. Still, as I raised Koko, I tried to gauge her performance by every available yardstick, and this meant administering infant IQ tests.

---

*Many higher intellectual functions still bear the hallmarks of their origins in a gesture. Gordon Hewes argues that as evolution progressed, the visual-gestural and vocal-auditory channels underwent a division of labor. Today, according to Hewes, we see the visual-gestural channel operating in higher mathematics, hard sciences, and technology "in the familiar forms of algebraic signs..., flow-charts, maps, symbolic logic, wiring on circuit diagrams, and all the other ways in which we represent complex variables, far beyond the capacity of the linear bursts of speech sounds." The vocal-auditory channel, on the other hand, operates in "close, interpersonal, face-to-face communication, in song, poetry, drama, religious ritual, or persuasive political discourse." In short, the visual-gestural remained the preferred mode in those areas where language and technology merge.

14

Testing Koko's Intelligence

The question people most frequently ask me about Koko is, How bright is she? The answer would seem to be a simple problem of administering intelligence tests. Determining Koko's IQ, however, is not so simple, for a variety of reasons having to do with the nature of the subject, the nature of intelligence tests, and, perhaps surprisingly, the nature of intelligence itself.

Isolating and testing language use or intelligence in any creature is difficult, but Koko, because she is a gorilla—and a willful one at that—added several levels of complexity of her own.

Answers that seem perfectly plausible to a gorilla must sometimes be scored as errors on standardized tests. For instance, the Kuhlmann-Anderson Test has two questions with a distinct human bias. One question directs the child to "Point to the two things that are good to eat." The choices are a block, an apple, a shoe, a flower, and an ice-cream sundae. Koko picked the apple and the flower. Another question asked the child to point out where it would run to shelter from the rain. The choices were a hat, a spoon, a tree, and a house. Koko sensibly chose the tree.

But overshadowing such ambiguities is that familiar gorilla stubbornness, which showed up from the beginning of the project. Early in the second year we gave Koko a series of tests in which she was asked to sort large plastic letters by color—to find all reds or yellows, for instance. Koko's performance on these sorting tests was curious. She showed marked and dramatic improvement up to the 90 percent correct plateau. Then her performance plummeted. This pattern has characterized many of her subsequent test performances as well. The most likely explanation is that Koko is interested in these various tasks until she masters whatever is involved to her satisfaction. Once she is satisfied, she has no further interest in pursuing the matter, and will gaily fail under the apparent assumption that if she does badly enough we will stop giving her that test.

This was the case with the ACLC test, in which Koko was asked to choose from among pictured alternatives the one that matched a spoken phrase. Interestingly, her level of accuracy was fairly consistent throughout the test even though it involved three distinct levels of difficulty. Something other than complexity seemed to be limiting her performance. In fact, as in the color-sorting tasks, there was a pattern to each session of testing. She would respond correctly to the first five or so questions and then begin to use one of her strategies to ruin the test and perhaps to show us that she had had enough for one day.

Koko has so far used ten distinct ways of failing, and she gives us no sign that she is running out of ideas. She will generally slip into this obstinacy when we have been administering a test for a long time or when we repeat questions. One of her favorite methods of failing a test that involves choosing from pictured alternatives is to stare at the right answer and point to the wrong one. To catch her at this game we are trying to figure a way to rig a video camera to record whether she looks left, center, or right so that we can match this recording with the answer sheet. Another of her favorite methods is to pick one position—for instance, upper left if she is being shown four answers in a square—and point consistently at that position no matter what question she is asked. Or she'll point to an answer immediately, without looking at the alternatives. Sometimes she keeps tapping a picture in vocabulary tests as though the answer is on the tip of her fingers but she can't quite remember it. Other times she ignores the question and asks for the bribe: Question: "What swims?" Answer: *Gimme candy.* Another method is to ask me to tell her the answer: "Who is this?" I ask (pointing at Eugene Linden). She signs, *Tell me.* Yet another is to say she doesn't know, even if the question asks her about her favorite sign or object. Finally, she'll simply say she is tired.

It seems to me that when Koko points consistently to the same position, or taps pictures as though summoning an answer to her brain, something more than stubbornness may be inhibiting her. Of course, the question could simply be too hard for her. Once, however, we gave her a vocabulary test that she breezed right through. Three months later, under exactly the same conditions, we administered the same test, and Koko tapped pictures distractedly throughout—seemingly willing to try, but unable to produce.

It is abundantly clear, however, when Koko's performance is suffering from a fit of obstinacy. She has more demonstrative ways of signaling her displeasure with testing. She will turn off the videotape camera or tape recorder, or she will run away. Finally, and quite often, she will simply fold her arms and stare at her navel. It's enough to make one throw up one's hands and agree with assertions about the gorilla's "limitations."

We are engaged in a never-ending search to find ways of administering tests to Koko without having the results skewed by gorilla static. First, it is necessary to fathom the reasons for this stubbornness. One reason may be that gorillas do not like to be told what to do. Another is that they rarely like to repeat things for no evident purpose. And a third reason is that Koko, at least, seems to derive some amusement from seeing us frustrated. Not much can be done about the first two reasons, but we are trying to figure some way to deal with the third. If Koko did not think that she was frustrating us, she might cooperate better during testing.

When testing Koko's intelligence, we followed the rules of the test. If it was to be given in one session, we did so. If there was no time limit on the test, we accommodated Koko's impatience by breaking up the test into several sessions. This probably did not give Koko any real advantage, since the questions in these tests—like "put the button in the box" or "show me what we drink out of"—were not the sort with answers that she might determine through a lot of homework. Koko's IQ testing did differ from that of a human child's in that she might be given the same test more than once. I did this because I wanted to get repeated measures as her development progressed. In Koko's case familiarity bred not increased scores, but contempt. Her scores often dropped on a test that was repeated, as boredom seemed to become a factor. Nonetheless, her scores were remarkable.

From September 1972, when we adminstered the Cattell Infant Intelligence Scale, through May 1977, when I administered form B of the Peabody Picture Vocabulary Test, she has scored consistently in the 70 to 90 range on different IQ scales. These scores reflect her mental age divided by her chronological age, the result of which is then multiplied by 100. Such scores in human infants would suggest the subject is slow, but not mentally retarded. In Koko's case, it is specious to compare her IQ directly with that of a human infant. For one thing, the two mature differently. Many of the early tests require mostly motor responses. Gorillas develop locomotor facilities earlier than human infants do, but, of course, do not develop bipedal walking skills or fine motor control as quickly or as well as we do. Moreover, as suggested earlier, there is not an exact match between the level of maturity of a three-and-one-half-year-old gorilla and a three-and-one-half-year-old human infant. Therefore, since chronological age is the divisor in the equation that is used to compute IQ, the IQ obtained is not very useful for comparative purposes.

What is significant is the steady growth in Koko's mental age, particularly in the tests that involve problem solving rather than motor facility. In this same period between September 1972 and May 1977, Koko's mental age grew in human terms from 10.8 months to 4 years 8 months. At age five-and-one-half, she had about the mental age of a human child of 4 years 8 months. Looking at this another way, in a sample 22-month period in 1975–76, Koko's mental age increased 19 months. Her mental age, measured in human terms, was not increasing at the same rate that she grew older, but it was not lagging too far behind.

In some types of questions Koko did better than human counterparts of her age. At age four-and-one-half, she scored better than the average child of six in her ability to discriminate between same and different, and in her ability to detect flaws in a series of incomplete or distorted drawings. She astonished me with her ability to complete logical progressions like the

Ravens Progressive Matrices test. Shown the following two problems from a similar test, she pointed to the correct answers (unknown to the tester) with almost no hesitation:

Which of the four lower selections best completes the series on the top?

Which of the lower boxes best completes the series on the top?

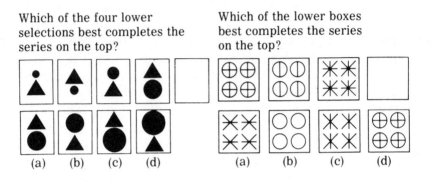

   (a)   (b)   (c)   (d)   (a)   (b)   (c)   (d)

Koko generally performed worse than children when a verbal rather than a pointing response was required. When tasks involved detailed drawings, such as penciling a path through a maze, or precise coordination, such as fitting puzzle pieces together, Koko's performance was distinctly inferior to that of children.

In the information part of the Wechsler Preschool and Primary Scale of Intelligence, one of the questions asked Koko to name two animals. She named the cow and the gorilla. And when asked to name what lives in water, Koko replied, *Tadpole good.* This test also contained picture-completion problems, with test cards showing drawings of objects with a part missing: a hand without a fingernail, a cat with one set of whiskers. Koko at age four whizzed through this at the level of a six-year-old human.

Over the years other apes have been administered intelligence tests. An orang once reportedly scored about 200 on an infant intelligence test, a result that may have said more about the orang's faster-maturing motor control than its reasoning abilities, although there is no question that orangs are bright. Viki, the female chimp who was the subject of an early attempt to teach spoken language to an ape, was given a number of intelligence tests and performed quite well. She did better than a control group of human infants up to the age of eighteen months, and matched their performance until about age three. Because Viki never learned to utter more than five or six words, she could not take those parts of intelligence tests that required language.

The fact that Koko was successfully learning a human language permitted me to test her intelligence in a much more thorough and ongoing way than has been possible with any other ape. And, of course, Koko's performance has surpassed that of any other ape. This is not because Koko is a genius among gorillas. It is simply because we have given Koko a tool, sign language, which allows access to her native intelligence.

It is hard to draw any firm conclusions about the gorilla's intelligence as compared to that of the human child. Because infant intelligence tests have so much to do with motor control, results tend to get skewed. Gorillas and chimps seem to gain general control over their bodies earlier than humans, although ultimately children far outpace both in the fine coordination required in drawing or writing. In problems involving more abstract reasoning, Koko, when she is willing to play the game, is capable of solving relatively complex problems. If nothing else, the increase in Koko's mental age shows that she is capable of understanding a number of the principles that are the foundation of what we call abstract thought.

This might seem an unsatisfactory answer to the question of Koko's intelligence. But one of the historical ironies that has accompanied our slowly developing picture of ape mastery of human abilities has been the disintegration of our picture of what these human abilities are. Just as we now acknowledge that speech is not the only possible mode for language, it is becoming clear that what we commonly test and discuss as "intelligence" is really a collection of abilities that perhaps should not be lumped together.

While on the one hand some psychologists believe that vocabulary testing is the single most reliable indicator of intelligence (or at least of performance during schooling), other scientists argue that abstract intelligence is basically not verbal, but pictorial, although many of its properties may be expressed verbally. This fragmentation of our image of intelligence has followed relatively recent developments in neurology which have suggested a greater degree of compartmentalization in the human brain than was previously believed.

The problem with answering the question "How bright is Koko?" turns out to be the problem of what we mean by the word *bright*. Intelligence tests, especially those for young children, ride roughshod over the various abilities we lump together as "brightness." Roger Fouts has said that if we tested chimps as interpreters of social relationships, they would far surpass humans. I'd add gorillas to that statement.

What we can draw out of the morass is Koko's ability to demonstrate some of the different facilities involved in intelligence. The tests, particularly the Stanford-Binet, show a capacity for analogic reasoning, and in this capacity—the ability to see the ways in which something is like something else—is the foundation of the disparate mental operations we call "intelligence."

We can say that Koko shares many of our mental experiences. When she looks at the world, she is conscious of her surroundings and, to a degree, of the laws that govern them in much the same way that we are conscious.

If intelligence testing gives an unsatisfactory picture of the scope of Koko's abilities, there is dramatic evidence of the sophistication of her thinking in the extraordinary ways in which she has used language. And, of course, determining the ways in which she uses language is the principal business of Project Koko.

PART III

The Gorilla View of Things

15

Koko's World

When Koko was two she was given a little rubber gorilla doll. At first she was afraid of the doll; she avoided it and would retreat when one of my assistants approached her with it. But after Ron had tickled her and put her in a good mood, she allowed him to place the doll on her stomach. Instead of throwing the doll off, as Ron expected she might do, she picked it up very gently. First she tucked it under her arm and walked slowly around the room. Then she briefly carried it pressed to her bottom. Finally, she settled on carrying it with one hand between her thigh and stomach and walked around on her free leg and arm. In the wild, a gorilla habitually carries her baby under her stomach (as Koko ultimately carried the doll), and then, once the infant has become a little stronger, clinging to her back. Of course, as a practical consideration, Koko couldn't keep the doll on her back. (When Koko was an infant, her mother had made her ride on her back only a month after birth, and Koko had to struggle not to fall off. Koko now often tries to balance her dolls on her back as her mother carried her.)

Vying with Koko's maternal instincts was her infant nature, and she treated the doll alternately with extreme tenderness and extreme roughness. She'd gently place the doll on the floor, and then jump on it. Or she would dangle the doll playfully above her, and then a moment later try to pull an arm off. In short, the doll got much the same treatment as would any two-year-old's.

Throughout her life, Koko has had a variety of dolls, stuffed animals, and pictures of animals to play with. While she was at the zoo nursery, I could point out the objects, people, and animals that paraded by and name them for Koko, but we had to teach her the great preponderance of her signs by showing her referents that were pictures or toys. Almost from the beginning of the experiment, Koko was forced to distinguish between representation and reality. The first few times Koko saw pictures of food in a book, she tried to eat the pages. But quite early she came to understand that pictures stood for something else not present. Indeed, the very constraints of her upbringing may have helped in this process of analyzing her world.

One of the most obvious things about Koko's environment was that she was walled off from the rest of the world, first by the glass of the zoo nursery, and then by the walls of the trailer. Quite naturally, Koko wanted to go out and play, but, as she discovered early in life, being able to go out and play was a treat accompanied by what probably seemed to her elaborate preparations. Consequently, Koko had to get accustomed to living a life somewhat detached from the panorama of events she could see beyond the window. This constraint led to one of her earliest linguistic inventions, and also to the first instance in which Koko talked to herself.

One of the first signs I repeatedly showed Koko was *bird*, a sign made by placing the index finger and thumb together in front of the mouth as a representation of a bird's beak. From the beginning, Koko has not liked birds. She hated the screeching of the raucous jays that alit outside her window. *Bird*, in fact, was one of the first words Koko adapted to form an insult. On this day in October 1972—three months into Project Koko—I took advantage of the fact that a turkey had settled outside the nursery window to show Koko once again the sign for *bird*. The bird was just on the other side of the glass, and this time when I made the sign, Koko watched my hands intently. Then she made a sign like *bird*, but with her fingers held away from her mouth, on the glass at the spot where the turkey was outside. Before and during this period of her training, Koko had been doing a fair amount of manual babbling—experimenting with different configurations of her hands. Now, however, she moved her hands away from the glass, and in a far more deliberate way than the rhythmic and carefree manner of her babbling, began to make this adapted *bird* sign on various objects. She seemed to be taking inventory of the different possibilities of her new word, and the objects she touched gave clues to the meaning of her adapted sign. She made the sign on one of the boxes she'd stack up to get to objects that were placed out of reach; she made the sign on a toy post-office box that she also used as a stand; she made the sign on a book I was holding above her head and out of reach; and, finally, she made the sign on the window that separated her from the outside world. She did all of this without prompting of any type; it was something she seemed to be thinking through herself.

If much of what fascinated Koko was out of reach, she had a means of access to this world that the other animals in the zoo did not have. This was language. Even in the wild, gorillas do not grow up in carousing groups of youngsters as chimps do, but in small bands and family units. Koko faced a different type of solitude. She suffered from no lack of attentive companions, and her environment was certainly enriched, but for all these controlled stimuli, she grew up somewhat insulated from the rough-and-tumble real world. Her world was hyperborean; she was safe from the terrors of the forest. This might have helped free any latent powers of speculation and analysis Koko had, but it did not prevent her from creating imaginary terrors from among the objects with which she was presented. First among these was the dreaded alligator.

Koko has never seen a live alligator. She has seen my pet iguana, named McGuana. The iguana is a harmless lizard, and of this innocuous species, McGuana is one of the more comatose, unintimidating specimens. Koko is terrified of him. She always has been terrified of him. I wish there were some way to let McGuana know this—it might do worlds for his ego—but on those rare occasions when he has encountered Koko, McGuana has been preoccupied with his own fears. McGuana once saved the day for one of my assistants, Candy Davidson. Koko had tricked Candy into letting her out of her room, and then refused to go back in. She banged on the walls, banged on Candy, helped herself copiously to the food in the refrigerator, and generally did everything to show who was boss. Ann Southcombe, in the other part of the trailer with Michael, had the presence of mind to call another assistant and ask him to fetch McGuana from the house. The sight of the approaching lizard was enough to cause Koko to flee back into her room and hide in the corner.

If the reality upon which Koko built her imaginary fears was only a mild-mannered iguana, she was more than equal to investing these fears with numinous intensity. At first she was afraid of any representation of an alligator—plastic, leather, or stuffed. As she grew older and more sophisticated, Koko became less afraid of the rubber alligators, but retained her fear of the stuffed and realistic-looking alligator toys. Koko is probably afraid of the teeth, because she has no fear of damaged toy alligators with their lower jaws missing.

In recent years I have exploited her fear of alligators to keep Koko away from parts of the trailer that are fragile or where she might get hurt or create a mess. At one point I had tacked up so many rubber alligators in Koko's trailer that at first glance one might have mistaken the trailer for the spirit house of some ancient religion. When asked what she is afraid of, Koko will often answer, *Alligator*. Once she was asked, "What do you do with bad alligator?" to which she responded, *Gorilla afraid*.

On many occasions Koko enjoys being terrified by alligators in much the same manner that children enjoy being terrified by ghost stories and horror films. She likes it when I chase her around the room with an alligator. She turns the game around as well. More than once she has grabbed a rubber snake and chased Ron around the room, actually making the snake "bite" Ron on the arm. Ron fulfills his part of the charade by pretending to be frightened.

Koko also seems to have discovered that she can threaten verbally with alligators. Once she was waiting impatiently for assistant Cindy Duggan to prepare her food. *Alligator chase lip*, Koko signed (she uses *lip* for any female). Cindy, perplexed, asked, "Alligator?" Koko signed, *Alligator do that hurry*, pointing to the plate of food. The astonished Cindy noted in the daily diary: "I think she's threatening me with an alligator if I don't hurry with the food."

There is one alligator Koko is not afraid of, and that is a red plastic puppet alligator. Koko spends a good deal of time talking to and playing games with her toys. This is a private pastime of hers, and she does not like to be watched when she is doing it.

She seems to get embarrassed when she discovers someone observing her while engaged in such play, and she abruptly breaks off whatever she is doing. One day I noticed Koko signing *kiss* to her alligator puppet. When she saw that I was looking, she abruptly stopped signing and turned away. On another occasion when she was five, she was playing with her blue and pink gorilla dolls. First she signed *bad bad* to the pink doll and *kiss* to the blue one. Next she signed *chase tickle* and hit the two dolls together. Then she joined in a play wrestling match with both dolls. When it ended she signed *good gorilla good good*. Finally she looked up and saw she was being watched, and abruptly stopped.

Another of Koko's pastimes is drawing. I don't think Koko is particularly artistic. However, she can come up with some fair representations, especially if she is copying from a picture or model. She uses appropriate colors and gets objects in their correct places. When drawing from her imagination, Koko's favorite subjects are birds and alligators. Even if her artwork is not strictly representational, Koko will enlighten us by telling what she has drawn when asked. Koko, by the way, is left-handed, and can use the precision grip when holding pencil or crayon.

Koko has occasionally grabbed a pen and exuberantly filled out the checklist we keep to record her daily signing behavior. She does this by drawing little circles as we do in the space beside each word. Koko's circles, drawn with a counterclockwise motion, are not perfect—they are more like triangles. She has also made crude stabs at copying our printing.

Another of Koko's solitary activities is talking to herself. Human children, when they are first learning language, spend much time talking with themselves. One sample of one-year-olds in nursery schools showed that about 60 percent of the infants' utterances were self-directed. By four, when children are more responsive to the people around them, this figure dropped to 27 percent of their verbalizations. Koko's percentage was never that high. Still, she does sign to herself, particularly when she is playing alone. The number of her self-directed utterances increased from about one every five hours at the beginning of the project in 1972–73 to about six per hour in 1976. The length of these utterances increased from about one word to an average of two words. In one representative sample taken when Koko was five, signs made to herself accounted for about 7 percent of the total she made in an hour.

Talking to themselves gives both the child and the gorilla the chance to experiment with and learn about a powerful new tool. In all probability, it reinforces the development of those parts of the brain that are necessary to language. For Koko, talking to herself seems to have other functions as well. Besides enabling her to practice language, it seems to have some comforting effect. She often signs her favorite words, such as *red*, or makes signs about the parts of her face. Once, for instance, while looking in a mirror, she signed to herself, *Eye, teeth, lip, pimple*, seeming to take an inventory of her face, not unlike any adolescent. These are all signs Koko knows very well, so she is not using them for practice. Perhaps she finds it reassuring to make the familiar gestures while seeing the parts of herself that she is referring to.

The most plausible conjecture is that Koko signs to herself simply to comment or exclaim about something that arouses her interest. One day she made a nest for herself out of a blanket, and then casually picked up the objects she found in it. Rummaging around, she found a toy horse. She signed *listen*, put the horse to her mouth, and huffed as she does to her toy telephone. Next she pulled out a mukluk with a flower on it. She picked it up, sniffed it, and then signed, *That stink* (she calls a flower a *stink*). Shortly thereafter, she smelled the blanket. Again she signed, *That stink*. Koko is nothing if not fastidious.

This is all to say that there is a lot going on in Koko's world, even when she is not in active conversation with her human instructors. Many of her playtime activities have to do with exploring the potentials of the tools she has been given—signing as she leafs through magazines, for instance. Language proved useful to Koko, even from a very young age, in her attempt to understand the circumstances in which she found herself. It allowed her mind to flower in a number of ways I was not testing. Language was a necessary part of the world Koko was growing up in, but it was also useful in her own self-examinations and self-musings. And the circumstances of Koko's world fostered her development. So much of it was out of reach, accessible only through symbols or pictures or models.

16

Gorilla Humor and Other Language Games

Koko appreciates a good joke even if she does not utter it herself. In fact, Koko relishes a good joke even if it comes from someone about whom her feelings are mixed. Such a person is Ron Cohn, who has advanced Koko's cause since Project Koko began and yet does not receive the gratitude from Koko that he deserves. From Koko's infancy, Ron has had the unenviable task of "enforcer." He keeps Koko in line when there are people around whom Koko does not accept as dominant. Because there are now few people that Koko does accept as dominant, it is more important to maintain Koko's respect for Ron than to try and improve relations between the two of them. If Koko is aware of Ron's dominance, she also begrudges him his power and, in the fashion of those who believe they are unable to redress their wrongs physically, Koko gets back at Ron with words. When asked "Who's Ron?" Koko almost invariably replies something like *Stupid devil* or *Devil head*. Yet when asked "What is funny?" Koko once replied, *Koko love Ron*, and kissed Ron on the cheek. This statement was remarkable because at that time Koko almost never used Ron's name sign, preferring to draw on her lexicon of insults when referring to her sometimes stern stepfather.

Koko has made similar responses to the mention of Ron's name in pig latin. Asked who "Onny Ray" was, Koko replied, *Trouble devil*. (Asked what "andy cay" was, Koko replied *Koko candy*, indicating that pig latin was a code she had no trouble breaking.) We often asked Koko to free-associate on words— "What does _____ mean to you?"—and when we asked her what Ron meant to her, she signed, *Knock bite. Knock* is Koko's coined sign for the English word "obnoxious"; there is no sign corresponding exactly to *obnoxious* in Ameslan.

Still, when Ron makes a joke, Koko will generously laugh in appreciation. One day while giving Koko a few M&M's, Ron said, "And one for the alligator," as he placed an M&M in an inflated alligator's mouth. Koko chuckled heartily.

Many of Koko's language games involve homonyms and rhymes and are typical of the exploratory games that children use as they learn about spoken language. One of the first indications that Koko could use words on the basis of acoustic similarity occurred when she would substitute a sign for an English homonym of a word she did not know. For instance, when Koko had difficulty articulating *need* she would occasionally use *knee*, a sign that sounds like *need* but that is made in sign language in an entirely different manner. She has also on occasion interchanged signs for *I* and *eye*; *know* and *no*; *eleven* and *lemon*; and others.

Maureen Sheehan, one of Koko's teachers, has also worked with Tommy, the first autistic child reported to be taught sign language. She notes that both Koko and Tommy substitute a similar sounding but inappropriate word when the right word does not come to mind. For instance, both will use the body part *back* in phrases such as *come back*. On the other hand, when both Tommy and Koko know a word well they make no such mistakes. Both understand that *like* means "similar to" and "love," and they recognize the appropriate sense of the word from the context of the sentence; in their translation to sign of a spoken sentence like "This fruit is like that fruit," they will usually use the sign *same*.

Even when she knows a sign, Koko occasionally utters strings of signs that in English sound the same. When asked to identify broccoli once, she signed *flower stink fruit pink...fruit pink stink*. I said, "You're rhyming, neat!" Then, to my astonishment, Koko signed, *Love meat sweet*. Requests for identification of certain objects and activities produce from Koko such statements as *Apple pink drink* and *You lip sip.* While it is arguable that these are not games but Koko searching for the right word, this explanation is implausible because the rhymes were on words Koko has used correctly tens of thousands of times—and because the words rhyme in English, not Ameslan.

Moreover, since mid-1978, Koko has obligingly generated rhymes when asked. In August 1980, for instance, I asked Koko, "Can you do a rhyme?" Her response: *Hair bear.* I continued, "Another rhyme?" She answered *All ball.* We have tested her ability further with an animal game. Arranging toy animals in a row in front of Koko, we ask her questions about them:

BARBARA HILLER: Which animal rhymes with hat?
KOKO: *Cat.*
BARBARA: Which rhymes with big?
KOKO: *Pig there.* (She points to the pig.)
BARBARA: Which rhymes with hair?
KOKO: *That.* (She points to the bear.)
BARBARA: What is that?
KOKO: *Pig cat.*
BARBARA: Oh, come on.
KOKO: *Bear hair.*
BARBARA: Good girl. Which rhymes with goose?
KOKO: *Think that.* (Points to the moose.)

Koko can also make gestural rhymes by altering one of the four components of the gesture; most often, she'll keep the proper configuration and change the location of the sign. One recent afternoon, for instance, I asked Koko to do a gestural rhyme, first demonstrating by signing *nail* and *teeth. Nail* is made by tapping the fingernail repeatedly with a crooked index finger; *teeth* is made the same way, but on the teeth. Koko then responded, *time* and *bellybutton. Time* is made by repeatedly tapping the wrist with a crooked index finger; *bellybutton* is made with the crooked index finger on the navel. About ten minutes later, Koko spontaneously signed, *bread red head.*

Much of Koko's humor is stimulated by her desire to inject a little life into otherwise boring routines. On December 23, 1977, Cindy Duggan was tending Koko. At one point she picked up an

almost empty jelly jar. Perking up, Koko said, *Do food.* Cindy
asked, "Do where, in your mouth?" Koko replied, *Nose.*
"Nose?" asked Cindy.
Fake mouth, replied Koko, explaining herself.
"Where's your fake mouth?" asked a persistent Cindy.
Nose, replied Koko.
Koko reaffirmed this inspired response on several occasions
over the next few months with various teachers. On January 11,
1979, Barbara Hiller had the following conversation with Koko:

KOKO: *Thirsty drink nose.*
BARBARA: *Your nose thirsty?*
KOKO: *Thirsty.*
(Barbara gets out some apple juice.)
BARBARA: *Where* do you *want this?*
KOKO: *Nose.*
BARBARA: Okay.
KOKO: *Eye.*
BARBARA: Right *in your eye.*
KOKO: *Ear.*
BARBARA: Okay, here it goes *in your ear.*
(Koko is laughing.)
KOKO: *Drink.* (She opens her mouth.)
BARBARA: Okay, that's the *best place for* a *drink.*
(Koko laughs and walks over to the window and sits down.)

Koko also lets her imagination play fancifully on the images
created by different activities. Sometimes when she becomes
too persistent in her requests for prepared drinks she is given
water, which she drinks through a straw. One day, after
nagging Barbara Hiller for drinks all afternoon, Koko became
discouraged when Barbara told her she could not have another.
Koko signed, *Sad elephant.*

BARBARA: *What* do you *mean?*
KOKO: *Elephant.*
BARBARA: Are *you* a *sad elephant?*
KOKO: *Sad... elephant me... elephant love thirsty.*
BARBARA: I *thought you* were a *gorilla.*
KOKO: *Elephant gorilla thirsty.*

BARBARA: *Are you a gorilla or an elephant?*
KOKO: *Elephant me me... Time.*
BARBARA: *Time for what?*
KOKO: *Time know Coke elephant good me.*
BARBARA: *You want a drink, good elephant?*
KOKO: *Drink fruit.*

The subject of elephants had not arisen that afternoon, but Koko had been using a fat rubber tube to drink with from a pot of water on the floor, rather than her usual straw and glass. Barbara thought that this might possibly have reminded her of an elephant's trunk, so she showed the tube to Koko.

BARBARA: *What's this?*
KOKO: *That elephant stink.*
BARBARA: *Is that why you're an elephant?*
KOKO: *That nose.* (Points to the tube; then wanders away laughing and returns in a moment.)
KOKO: *That there.* (Points to a can of soda, then to her glass.)
BARBARA: *Who are you?*
KOKO: *Koko know elephant devil.*
BARBARA: *You're a devilish elephant?*
KOKO: *Good me thirsty.*

In this example, Koko is linking two unmodulated signs together to achieve—intended or not—humorous effect. The result is an absurdity—an elephant gorilla; a gorilla with a rubber straw trunk. Moreover, apparently Koko summoned the word *elephant* on the basis of the fancied similarity between the rubber straw and her memory of one distinguishing characteristic of elephants. There were no elephants present, of course. Thus Koko's phrase involved pure analogic thinking. The humor derived from the controlled incongruity of her analogy. It is the type of humor that is only possible through language, because it involves playing in a controlled way with the expectations of the listener. It is similar to the earliest jokes of children, which are also based on incongruity. And, like a child, Koko is her own best audience.

Koko's jokes, which play on the structure of Ameslan, require that Koko's audience recognize the sign behind her distortions.

Koko has distorted the sign *drink* in several ways for humorous and bratty effect. Once, Ron pointed to a picture of bottles on a wall and asked, "What's this?" Koko responded, *Drink funny there*, making the drink sign on her nose. This distortion produced a nose-thumbing gesture equivalent to the sign for *rotten*. Koko then made her intent clear by signing, *funny there*.

Koko's humor can also be remarkably unsophisticated, and if a prank gets a rise out of someone, Koko is apt to repeat it. After noting with pleasure that blowing an insect on me produced a shriek and a jump, Koko did it again, this time laughing. Koko's laugh is a chuckling sound that is like a suppressed, heaving human laugh.

Another of Koko's practical jokes is the attack with a plastic alligator. Koko will sneak up on some (supposedly) unsuspecting human with a little toy alligator hidden behind her back. When she is near her victim, Koko will abruptly spring up, brandishing the alligator wildly. The human is expected to assume a terrified look, scream, and run. Koko knows that the whole thing is a charade, but she thinks it is vastly funny. It is wonderful that a gorilla might even for a moment think that it needs a prop like a toy alligator in order to scare a human.

Other examples of Koko's humor blend into examples of contrariness: her insisting that a white blanket was red, and then producing as proof a minute speck of red lint; responding to a request to smile for a photograph by signing *sad-frown*. Humor, then, is another of the dividends from the gorilla's stubborn streak. It is one part of the pattern Koko spontaneously introduces to express her independence and enliven the dull routine of schooling.

Koko's humor is one of those subtleties that might be missed in what Roger Fouts calls an "overstructured situation." Her jokes make sense because they resonate with other aspects of her personality that she expresses through language. It is reasonable to assume that her jokes are intentional because she laughs and because her modulations and rhymes show that Koko has the capacity to manipulate the language in the ways demonstrated in her humor. However, humor, as is the case with many other of Koko's creative uses of language, is difficult to investigate through formal testing. So far, we have had to let Koko take the lead and decide when it is that she wants to say something funny.

The essence of Koko's humor is her ability to diverge from norms and expectations in a recognizably incongruous manner. Her humor taps in various ways her capacity for analogic thinking, and it also shows a capacity for displacement, a cardinal attribute of symbolic communication. These capacities are further confirmed by a number of Koko's other creative uses of language. Not all of these inventions are as engaging as humor.

17

Innovations and Insults

Below is a list of word combinations invented by hearing children, deaf children, and Koko to describe things presented to them. See if you can tell the difference between man and ape:

"barefoot head" to describe a bald man
"giraffe bird" to describe an ostrich
"fireplace wall shelf" to describe a mantelpiece
"elephant baby" to describe a Pinocchio doll
"finger bracelet" to describe a ring
"white tiger" to describe a toy zebra
"red corn drink" to describe pomegranate seeds
"bottle match" to describe a cigarette lighter
"eye hat" to describe a mask

The first two were the creation of hearing children; the third, a deaf child's; and the rest were Koko's. Koko is not an anomaly among apes in her ability to use inventions to describe the novel. Washoe described chewing tobacco as "smoke string food," Lucy called a radish a "cry-hurt food," and Lana, using her computer console, came up with the slightly awkward "apple which is orange" to describe an orange.

However, if Koko is not alone among apes in her ability to create these inventions, she is remarkably predisposed to purple descriptions even when she knows the proper word. For example, Koko provides evidence that "seeing red" crosses species lines. Several times she has described herself as *red*

Table 5. Examples of Compound Name Innovations Made by Koko
(October–November 1977)

Age	Name	Referent
6 years, 2 months	*Milk candy*	Rich thick tapioca pudding
	Stuck metal	Metal and a magnet stuck together
	Cold hard	Companion signed *frozen*, for a frozen berry
	Fruit lollipop	Frozen banana
	Lollipop food tree apple	Caramel apple on a stick
	Pick face	Tweezers
	Fruit red seeds	Pomegranate kernels
	My cold cup	Ice cream
	Dirty orange	Lemon (Koko doesn't like them)
6 years, 3 months	*Blanket white cold*	Rabbit-fur cape
	Lettuce grass	Sprig of parsley
	Nose fake	Mask
	Mask look	Stereo viewer
	Potato apple fruit	Pineapple (spoken)

rotten mad rather than simply mad. More than once she has said she is a *red mad gorilla,* which is enough to scare anyone.

Formal testing of Koko's metaphoric proclivities is, of course, a difficult problem. But, using a test of metaphoric matches devised by Howard Gardner, I have been able to collect some hard evidence suggesting that Koko can understand metaphor. Gardner uses a somewhat cumbersome definition of metaphor ("The ability to project in an appropriate manner sets of antonymous or 'polar' adjectives whose literal denotation within a domain...is known onto a domain where they are not ordinarily employed."); but the test is really quite simple. Once it is established that the subject knows the literal difference between such polar adjectives as light and dark, the test simply determines whether the subject's ideas about a color being happy or sad, hard or soft, and so forth, match the collective decision of a group of adults. The test is very long, and parts of it—such as a musical section—we couldn't give Koko because we didn't have the equipment on hand.

In those parts of the test we did give her, Koko performed better than both preschool children and the seven-year-olds tested by Gardner. Almost all of Koko's answers—90 percent— were matches with the norms set for the test. In contrast, seven-year-old children matched in 82 percent of their responses.

The idea that a color is happy or hard, or that "red mad" means very angry, requires even more sophisticated analogic thinking than is needed to see a nose as a fake mouth. It is based not on obvious feature similarities, such as that of a straw to an elephant's trunk. Rather, these metaphoric matches have to do with a common emotional resonance spurred by two entirely dissimilar words. Unfortunately, Koko is most eloquent in generating this type of metaphor when spurred by such emotions as anger, jealousy, impatience, and fear.

Once, for instance, I said to Koko, "I think Mike is smart. Is he smarter than you?" Koko's reply made her jealousy evident: *Think . . . Koko know Mike toilet.*

Typically, Koko's impatience drives her to metaphoric insults. The following conversation took place after Koko had been nagging Cindy for food and Cindy refused to comply:

KOKO: *Time nails nut.* (Koko appears to be calling Cindy a nut—one of her favorite insults—and threatening to scratch her with her nails—one of her favorite threats—if she does not acquiesce.)

KOKO: *Fruit . . . Key key time.* (Koko is suggesting that the time has come for Cindy to use her key to unlock the refrigerator.)

CINDY: *No, not now key time.*

KOKO: *Yes time come-on time, nut.*

CINDY: *No, not time!*

KOKO: *Yes time.*

CINDY: *No time.*

KOKO: *Nails.*

CINDY: *Why?*

KOKO: *Time.*

CINDY: Oh!

Koko also uses direct threats:

KOKO: *Bite big-trouble bite big-trouble.*

MARGIE: *What about bite big-trouble?*

KOKO: *You do apple hurry.*

MARGIE: *You don't do threats, no apple now, it's milk time.*

KOKO: *Do milk hurry.*

Table 6. Examples of Koko's Comments About the State of the Environment

Age	Utterance	Context
3 years, 6 months	Listen quiet	When alarm clock stopped ringing in the next room
	Nose stink	Perfume
	Listen drink listen	Sound of fizz from carbonated drink
	Smell stink	Cooked broccoli
	Cat that	Barbara's new leopard ring
4 years, 6 months	Baby	Mouse
	Moose	Moose pictured on stamp
	Pimple look	Companion had been looking at irritated spot on nose in mirror
5 years, 6 months	Cut tree	As companion cuts celery
	Bite there red	Koko indicates spot on her arm at the same place where she bit Mike the day before
	Cry Mike cry	To deaf assistant who is sweeping floor; Mike is crying
6 years, 6 months	Good know Mike	Mike finally said Koko's name after many wrong answers
	Hear bell	Oven timer rang
	Hug Koko gorilla	Drawing of two gorillas hugging
	Write bird there	Koko indicates her drawing
	That soft	Velvet hat
	Lady eyemakeup there	Picture of lady with makeup
	Nose funny	Penny: "Look at Ron." Penny laughs; Ron has toy spider ring on his nose
	That man	Picture of a gorilla

One day an enormous dump truck emptied its contents in front of the trailer, scaring the daylights out of Koko. Maureen asked her what was wrong:

KOKO: *Afraid...close drapes.* (Koko often asks to have the curtains closed when she's feeling insecure.)

KOKO: *Hurry do that.* (She points to dog-training prod. Maureen picks up the prod.)

KOKO: *You do alligator drapes red bellybutton.* (*Red bellybutton* is Koko's name for the prod, which has a red button. Apparently, she thought the truck would go away if Maureen threatened it.)

Table 7. Signs Koko Uses as Insults, Expletives, and Derogations
(Age 4 to 7)

Lexical Item	Context	Example of Use
Bird	Kate Mann won't let Koko open the refrigerator	Kate bird rotten
Darn	Prolonged chirping of bird in distress	Darn bird bird
	A friend Koko calls *Foot* leaves without saying goodbye to her	Darn Foot
Devil	Penny: "Do you like Mike?"	Devil rotten
	Penny: "Want Mike in? What think?"	Think stupid devil
Dirty	Penny: "What's wrong with milk?"	Dirty mouth dirty taste
	Penny: "What's this [lemon]?"	Dirty orange dirty lemon
False-fake	Koko is offered rotten grape	Mad fake fake
Frown	Kate asks "Why frown?" (Koko had earlier forgotten the sign for *pink*)	Frown that pink don't-know
Nut	Penny: "Are you jealous of Mike?"	Mike nut
	She sees Mike outside	Bad gorilla nut
Rotten-lousy	Bird chirping in distress outside	Rotten bird
	While pulling leash and hitting Mike	Rotten Mike
	Penny: "Learning is fun!"	Mad rotten
Stink (er)	Visitor asks "What's that?" about calendar photo of Koko's father	That rotten stink (er)
	Rubber monkey-face puppet	That stink (er)
Stubborn-donkey	At Cathy Ransom's request, Penny asks Koko to sign about drawings Cathy made to illustrate the numbers one and two	Know that stubborn-donkey
	Koko has repeatedly requested the drink Cindy Duggan is offering to a doll	Stubborn-donkey do that
Stupid	Penny playfully puts salt shaker on her head	Head stupid
	Mike has signed *Out out Koko Mike* to come out of his room to play with Koko	Know Mike devil stupid out
Toilet	Penny: "Say *bad*."	Rotten toilet
Trouble	After a scolding for pulling up flowering plants	Trouble flowers
	As Koko is scolded for pinching	Bad trouble
	Mike is about to come in and play	Trouble devil

The morning after Koko had received a severe scolding for ripping up a chunk of her wall-to-wall carpeting, she greeted me by signing *frown*. "Why frown?" I asked. She answered, *Penny dirty toilet devil.*

Koko has invented other insults, which she uses with abandon. She knows *nut* as food, but she also knows that the word is used during affectionate chiding for crazy behavior. Originally, when Koko used the word *nut* in this manner she would modulate the sign so that it was made with the fist at the side of her mouth rather than in the proper position in the middle. After we caught on that this use of *nut* was intentional, Koko gradually began making the sign the same way whichever meaning she intended.

Bird is a word that Koko decided had useful properties as an insult without any prompting by humans. She uses *bird* the way we might use "rat" or "birdbrain." As in the case with *nut*, at first Koko altered the sign when she was using it as an insult, but gradually eliminated her modulation as it became clear that we understood her intent.

As in her other creative uses of language, Koko has several different means to achieve her ends. She modulates the language itself, as in her adaptations of *bird* and *nut*; she exploits metaphoric associations of such words as *red*; and she links whole words to create compound names and to express complex associations. As in all the various modulations of sign language discussed in the past two chapters, Koko's metaphors and invective reveal that she is actively involved with the language, and has, on her own, extracted a number of rules to govern the ways in which Ameslan might be modulated. In many cases her discoveries match those made by children when they begin to play with this immensely powerful tool.

As I have observed Michael, the young male gorilla we acquired in 1976, struggle to find *le mot juste*, I remind myself that Koko has not always been as facile with sign language as she is now. Our ambitions for Michael go far beyond obtaining comparative data on language acquisition. Somewhere down the road we hope that Michael will develop sufficient charm to be a suitable language-using mate for Koko. In the meantime, despite such epithets as *stupid toilet*, and a rocky start to their relationship, Koko treasures Michael as her sturdiest playmate, and pupil.

With typical gorilla contrariness, Koko, asked to smile for the camera, responds by signing *sad-frown*.

Koko adds complexity to her signed utterances by modulating them. Here, the camera catches her partway through signing *hurry pour there* and *drink*. She shakes her right hand for *hurry*; signs a modified *pour* (made by turning the thumb of a fisted hand down); indicates *there* by pointing to the cup with her thumb; and simultaneously signs *drink* with her left hand.

Koko combines the sign *darn*—hitting the back of a fisted hand on a surface—with a natural gorilla expression of annoyance, lips pulled tight over teeth.

While Penny holds a doll, Koko makes the composite sign *Koko-baby*. *Koko* is made with the right hand on the left shoulder; *baby* is made with hands on elbows in a cradling gesture.

As she tickles Penny's foot, Koko laughs heartily.

Koko does full justice to the sign *sad*.

A close-up of Koko. She is holding a live bug gently between her lips.

Family portrait: Penny, Koko, and Ron.

When Penny asks Koko to ''teach'' her baby, Koko molds the chimpanzee doll into a sign resembling *eat*.

In play, Koko picks up a toy banana, signs *toothbrush* (left), and then uses it to ''brush'' her teeth (right). Piaget observed young children performing similar mental transformations on objects in play.

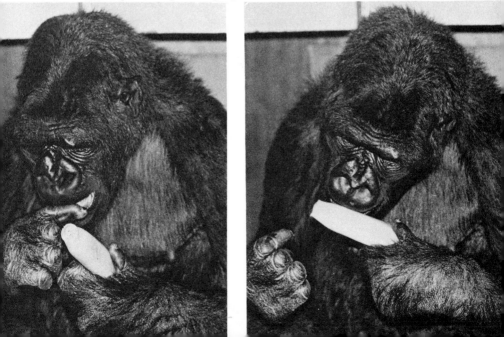

Chatting with herself, Koko signs *eye* while looking at a picture of a big-eyed frog.

Unjustly scolded for breaking a doll that Michael had actually half-destroyed, Koko retaliates by signing: *You...*

dirty...

bad...

toilet.

Koko signs *visit* to have Michael into her room to play. The sign is made with fingers in a V on the shoulders, but the gorillas use all fingers on their shoulders.

Koko bestows a kiss on Michael, who peers out of the elephant toy box.

Although Michael (on top) is two years younger than Koko, he's more than her match in their daily wrestling bouts.

Michael (in the middle) and Koko learn about a tomato worm, as Penny shows them the creature and signs *worm*.

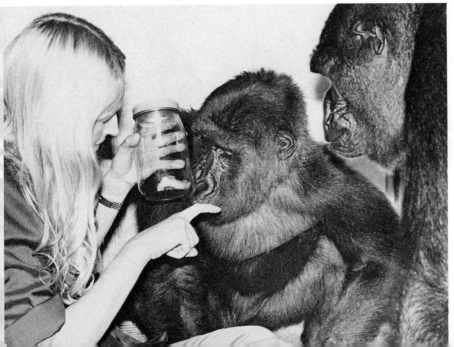

Michael gives Ron a toothy hug.

Koko kisses Michael, who appears to be moved by this gesture of affection. Penny hopes that the two will eventually mate.

Time for a tickle. Michael's teacher Barbara Weller breaks up his schooling routine with frequent play sessions.

In a humorous attempt to eliminate the possibility of cueing, Penny dons a mask and mirrored sunglasses and places her hands behind her back. Koko signs *bite* to Penny, whose mask has prominent pointed teeth.

Then Koko signs her combination *Koko-love* to her disguised and peculiar-acting "mother."

When Penny turns to get a shot of her mask, Koko mugs for the camera.

The trailer at its new site in Woodside.

The new house in Woodside is on seven acres of
secluded land, where Koko and Michael can romp in a
natural setting.

18

Koko and Michael

It is a basic fact about gorillas that they get very large and very strong. The trailer that promised release from the zoo when Koko was an infant now houses Michael as well, who in a few years will be far larger than Koko.

During his first few years with us, Michael occupied what used to be Koko's bedroom. Michael has his own small kitchen in his wing of the trailer, separated from Koko's quarters by a solid wooden door. Michael also has a small den, which is separated from Koko's room by a chain-link screen. This room has two doors, one that connects with Michael's room, and another that gives Michael access to Koko's part of the trailer. We designed this configuration of rooms so that Koko and Michael might pursue their studies undistracted by each other's antics but still have ample opportunities to play together or in sight of each other.

This they do, although I purposely restrict the amount of time they are permitted together so that they do not develop a sibling bond that might preclude future mating. When they first met, such a danger was not a worry. Initially Koko was jealous of Michael and, being older and bigger, tried to bully the little fellow. Before Michael arrived Koko had been told many times that a baby gorilla was coming. When Koko first saw Michael, who was considerably smaller than she but had imposingly large hands and feet, and at 40 pounds was not baby-sized, she signed, *Wrong old.* This gorilla did not fit her idea of a baby.

They would start to play together until Koko would take advantage of the jocular rough-and-tumble to sneakily bite and scratch poor Mike. Having seen this treachery visited on humans, I could often tell when Koko was about to try one of her tricks and warn her not to be rough. Reproached, Koko turned into the soul of wounded innocence, and frequently she would find some tiny scratch to point to, apparently pretending that she was the wounded party or that Michael had started the fracas. After a few weeks of Koko's terror, Michael began to stand up for his rights. Koko was at first dumbfounded by the fury of the little fellow, and very quickly thereafter settled into a more amicable relationship with Michael. Today Michael, though younger, is stronger than Koko.

Michael's personality is markedly different from Koko's. Some of the differences have to do with the fact that Koko is female and Michael is male. Perhaps some have to do with the fact that Michael, the second gorilla, does not get quite as much attention as Koko. The stardom that humans have conferred on Koko has, if only slightly, affected her behavior. Because Koko was the first language-using gorilla, she has frequently been surrounded with humans hanging on her every word. Michael's youth has been unencumbered with this sense of specialness, although he too has always had human companions. Chimps and gorillas are extremely sensitive to social relationships, and Michael gets hurt if visitors don't stop by to say hello before going on to see Koko. Consequently, he has to try just slightly harder if he is to receive his share of praise and attention.

Michael is a very good student. He seems able to concentrate for longer periods than Koko did at his age, and he is extremely persistent when confronted with a puzzle. Often he is to be found intently examining his toys. He likes to put things together, which Koko rarely does—she prefers taking them apart. While the two gorillas were out on a walk one day, we came across a sewer opening. Koko opened the cement lid; Michael closed it. Koko opened it again, and Michael closed it again. This went on until their human companions got impatient and ended the game.

On the other hand, for all his studiousness, Michael is still far behind Koko in vocabulary size and grasp of the language. Koko did have a considerable head start on Michael, since he was

three-and-a-half before he received his first tutoring in
Ameslan. However, despite the differences in their background, Michael serves as a constant reminder of how far Koko
has progressed, and makes many of the same mistakes Koko
used to make.

Michael acquired signs slowly at first; during the first year 19
signs met with my criteria and 17 met the Gardners'. This was
somewhat better than Koko did in her first year, but Michael
was older and we had the benefit of our experiences with her.
From the beginning, we selected a few basic signs from Koko's
early vocabulary and reviewed and molded them daily. Ann
Southcombe, who left her seven young gorilla charges at the
Cincinnati Zoo to help me, worked all day every day with
Michael—something I was unable to do with Koko in the early
days of the project. Ann's persistence and dedication contributed in no small way to the rapid development of Michael's
vocabulary.

Michael seemed to use his signs more spontaneously than had
Koko in the early days of the project, at least in the presence of
myself and Koko. He would comment on our activities, simply
as a spectator, with no external incentive. One day as he
watched me struggle with a jammed door, he signed *out out*.
What impressed me most about him in the early months was
that he actually tried to instruct some of the novice volunteers in
sign. As I was working with Koko, one volunteer called me to
ask what it meant to hit two fisted hands together. I told her
chase and asked why she had inquired. She said that Michael had
been signing this to her; when she repeatedly failed to respond,
he took her hands and hit them together, then gave her a push to
get her moving.

Michael tends to be more inquisitive than Koko, and his
signing reflects this. Michael forms questions through inflection in the same manner as Koko and the deaf, but he has also
begun to use the "wh" word "what" more frequently than Koko
ever has, and in a variety of ways. He poses direct questions,
such as signing *What good there?* while pointing to Barbara
Weller's backpack. He also seems to use the word to feign
surprise or innocence, as when he signed *what?* in response to
the statement *You a thief.* Sometimes it is difficult to figure out
what he means: *Do do do what do do*, which he signed while two

of his teachers were talking to each other, remains a mystery. But occasionally his motives seem clear, as when he was told he'd already received his banana, and he answered, *What banana?* (Actually, his bewilderment may have been real rather than assumed, since the banana he was given was not fresh but frozen.)

Despite his more frequent use of "wh" words, Michael was in general much less conversational than Koko early on. He tended to sign only about play and feeding and in response to simple questions during recorded sessions. Possibly this was due to the fact that most of his teachers and companions were volunteers not fluent in sign.

In August 1979 Michael seemed to break through a plateau in his language acquisition. He began signing more frequently and using a wider range of signs. In early June 1979 Michael's vocabulary had been 83 qualified words by my criteria, 59 by the Gardners' criteria. By late October 1979, Michael's vocabulary had jumped to 119 words by my criteria and 110 by the Gardners'.

One thing this increase indicates is the effects that a change in personnel or a bad relationship with a teacher can have on an animal's motivation to learn and to communicate. In December 1978 I hired a new teacher to replace Ann Southcombe, who had returned to Ohio. The new teacher never developed the rapport necessary to work effectively with Michael. Michael's vocabulary decreased sharply at his arrival and then recovered somewhat, jumping again after we replaced this teacher with another who related better to Mike.

One heartening aspect of Michael's efflorescence is that he is rediscovering many of Koko's innovations and creative uses of language. He has begun to shower his human companions with stories about events of the past. On September 30, 1979, Barbara Weller arrived to work with Michael and found him looking out the window signing *necklace*, the word for his leash.

BARBARA: *Necklace?*
MICHAEL: *Girl.* (Stares out the window again.)
(A few minutes go by.)
MICHAEL: *Know hit-in-mouth.*
BARBARA: *Know hit-in-mouth?*

MICHAEL: *Hit-in-mouth red bite.*
BARBARA: *Why* are *you signing hit-in-mouth?*
MICHAEL: *Know.*
BARBARA: *Who* do *you want* to *hit-in-mouth?*
MICHAEL: *Hair girl red.*
BARBARA: *What? Red hair girl?*
MICHAEL: *Lip.* (A sign both Koko and Michael use to refer to a woman.)
BARBARA: *Lip?*
MICHAEL: *Lip lip lip big-trouble.* (He returns to the window again.)

Barbara later discovered that before she had arrived there had been a terrible row across the yard at the Auditory Neurology Laboratory. A red-haired woman had stormed into the lab and gotten into a screaming argument with one of the research assistants. Apparently she had hit the assistant and was ultimately subdued at gunpoint by police. Koko and Michael had been glued to the front window throughout the incident, and had been upset by the turmoil.

Michael has also shown concern for the weak and helpless. For a period of time he spontaneously began signing about cats and birds. On August 1, 1979, Barbara asked him to identify a picture of a bird:

MICHAEL: *Cat mouth.*
BARBARA: What did you say? Does this go in a cat's mouth?
MICHAEL: *Eat.*
BARBARA: Do *cats eat this?*
MICHAEL: *Eat.*
BARBARA: Do *cats eat this?*
MICHAEL: *Cat me chase. Leg leg leg.*
BARBARA: *I don't-understand, what* did *you say?*
BARBARA: *What this?* (Picture of bird.)
MICHAEL: *Cat eat.*
(A few minutes pass.)
BARBARA: *What* did *you say about cats? What* do *cats eat?*
MICHAEL: *Bird.*

After several angry references to cats—*Cat cat chase eat bird...cat hit bad cat...cat bird bird frown hit-in-mouth*—we began to believe that Michael must have seen one of the wild cats inhabiting the animal facility yard catch and eat a bird outside the trailer. In June 1980, when Barbara asked Michael to tell her what *bird* meant, he launched into the following narrative: *Bird good cat chase eat red trouble cat eat bird.*

Like Koko, Michael has created metaphorlike combinations. He called peas *bean balls* and nectarine yogurt *orange flower sauce.* Michael also makes many of the same associations that Koko makes, and resorts to what appears to be humor to express his displeasure at what he deems to be unwarranted requests:

ESTHER ROBBINS: Can I have a *long sentence for meat?*
MICHAEL: *Meat meat meat meat meat meat meat meat meat meat.*
ESTHER: Well that's *long but* it's *silly.* Can you *do better?*
MICHAEL: *Do come-on gorilla like like.*
ESTHER: *What* do *you like?*
MICHAEL: *Do meat.*

Also like Koko, Michael occasionally uses signed threats when something is not done fast enough for his liking:

BARBARA WELLER: *Stop nagg*ing!
MICHAEL: *Hit hit nut come-on eat.*
BARBARA: *You threaten*ing *me again?*
MICHAEL: *Hit-in-mouth. Toilet toilet more nut there.*
BARBARA: *Toilet. Now you*'re *insult*ing *me! Where* did *you learn* to *nag, threaten,* and *insult?*
MICHAEL: *Eat nut.*
BARBARA: *Who taught you* to *nag?*
MICHAEL: *Gorilla lip.*
BARBARA: *What*'s the *gorilla lip*'s *name?*
MICHAEL: *Smile smile smile smile smile smile.*

Michael can trade insults with gorillas as well as humans. When Barbara told him that Koko had called him a *toilet devil,* Michael responded by signing, *Stink bad squash gorilla lip.*

One day, while Barbara was lecturing Mike about being too rough, Mike interrupted:

BARBARA: *...Can't play with rough gorilla*s. *You...*
MICHAEL: *Quiet girl.*

BARBARA: *... must* be *gentle.*
MICHAEL: *Quiet quiet.*
BARBARA: Are you telling me to shut up?
MICHAEL: *Smile.* (Laughs.)

Like Koko, Michael has been moved to lie from time to time. One day in May 1978, when asked by volunteer Ellen Strong, "Who ripped my jacket?" Mike signed *Koko.* Ellen, knowing that the culprit was the short hairy one looking at her so innocently, repeated the question. This time he placed the blame on me. Finally he confessed. It was *Mike.*

Koko has on occasion undertaken the job of coaching Mike in sign. Before they are allowed to play together, they must ask to do so in appropriate language. Sometimes in these situations, the pressure gets to Mike and he forgets such obvious words as Koko's name. In 1978, Michael drew a blank on Koko's name after signing *out.* Ann Southcombe, who was then working with Michael, refused to let him through. Koko, on the other side of the screen, began to get impatient. She signed to the puzzled Michael, *Do visit Mike hurry, Mike think hurry,* imploring him to come up with the right sign. Then she said, *Koko good hug,* and it finally dawned on Mike to say *Koko.* A relieved Koko congratulated Mike—*Good know Mike*—and then signed, *In Mike.*

Once they are together, their play is incredibly energetic, rough and loud, but it is still play. To the visitor outside the trailer, it sounds as if there is a small war going on inside, with all their pounding on the walls, tumbling, and rocketing about.

A typical session* goes like this: Koko and Michael are let in together and Ron leaves to get a cup of coffee. Michael at first looks out the window to follow Ron's departure, and then breaks off his vigil to grab Koko's leg. He starts chewing on her leg. Koko finds this ticklish and starts laughing. Mike stands up, and they begin wrestling. Koko starts biting his hand, and then kicks him away. Michael returns to look out the window, and Koko adjourns to her toilet. Seeing someone with a child outside the window, Koko claps, and then runs off to find Mike. I hear a crash coming from Michael's room, and find them in there wrestling. Koko breaks away, but Michael persists, and they

---

*This sequence is transcribed from a scoring taken January 1, 1979.

wrestle in the doorway. Koko glowers at Mike and then tries to run off. Michael grabs her and is dragged in turn for a bit. Koko ends up at the door to Michael's room. She goes in, and Michael comes out.

A little later Ron returns with treats. Koko and Michael spy Ron's return. "Ronny's here," I announce. "Has he brought something?"

Koko looks at Michael, who comes over to start playing, but Koko breaks away to continue to watch Ron's approach. Ron is carrying a number of things. He brings balloons to the front porch. "Koko, look, what is that?" Michael does a forward roll. "Koko, what is that?" I repeat. She signs *sip*; Ron is in fact carrying his coffee as well as the balloons. Koko then signs *straw*, another item Ron is carrying.

Ron enters and begins unloading his treats. I ask, "What's that?" Koko signs *good.*

"What's Ronny getting out now?"

Koko signs *drink*, but Ron produces a hamburger and garnishes.

Koko runs back and starts pounding Michael, "Hey, listen, Koko! That's not going to get you any food. I don't want you to hit Michael. You've got to be quiet. Stop it!" Koko stops hitting Michael.

I give Koko and Michael hamburgers. Koko signs, *Nice good drink.* I ask her to identify the hamburgers. *Bread sandwich,* signs Koko. Then I turn to Michael.

"Okay, we'll see what Michael says."

Me eat, responds Michael.

"What else? What is that, Michael? What is this? He knows sandwich, doesn't he? What's that?"

Nut, signs Michael. (He associates sandwiches with nuts, since he almost always gets peanut butter.)

"No. What is that? It's a sandwich, Mike."

Michael begins shredding his hamburger into thousands of pieces.

And so a typical afternoon continues, with continual roughhousing, cautions not to play too violently, and occasional conversations.

At present Koko signs much more frequently to Michael than he does to her. Michael's favorite signs when speaking to Koko are *chase, tickle,* and *Koko.* It is interesting that Michael sometimes addresses Koko with signs that he could only have learned by imitating her. Michael uses Koko's invented *lip* for woman and *obnoxious* sign, for instance. Even more intriguing is his variation of the *tickle* sign depending on whom he is conversing with. Michael learned to say *tickle* by drawing a forefinger across the back of the hand, and he uses this form when playing with humans. But when he is with Koko, Michael often uses her variation of *tickle,* which is made on her underarm. Perhaps Koko and Michael will develop their own gorilla jargon when using sign between themselves.

Koko has been experiencing estrus periods for about four years now, and to our relief she seems to be developing an interest in Michael as a potential mate. Over the past few years Koko has developed several crushes on men. One of the most memorable was a deep interest in a workman who was helping to build a laboratory across the yard from Koko's trailer when it was on the Stanford campus. This man, whom Koko nicknamed *Foot* (a word she came to use for all men), knew some sign language because his grandmother was deaf and spoke Ameslan. He was quite good-natured about Koko's coy flirtations, and on occasion Foot would share his lunch with Koko, even going so far as to bring two straws for his pint of milk. On her part, Koko maintained a near-constant vigil by the trailer window during the periods when he was likely to come into view. When she saw him, she'd blow him kisses, put decorative things on her head, sign *Foot,* and beg to go out and visit.

Koko also developed an interest in Dave Stallcup, who worked with Michael. This crush was more of a problem: Koko has a jealous streak, and she would make life miserable for any female volunteer she saw going into Michael's part of the trailer while Dave was there. She'd ram the door separating the two parts of the trailer and in every way possible suggest that the woman in question keep her distance from Dave.

Thus it is in the interest of domestic tranquility as well as science that we encourage Koko's relationship with Michael. And, indeed, Koko seems to be looking at Michael in a new light when she comes into estrus these days. Michael, on the other hand, still does not have a clue as to his role in this unfolding drama. It will be a year or so before Michael, now about eight-and-one-half, is old enough to respond to Koko's advances. Even now, though, Michael does feel protective toward Koko. He barks loudly when he sees male strangers touch Koko in play. Seeing a new volunteer untangling a leash that had me bound to (and dragged by) Koko, Michael mistook the man's lunge toward Koko to release us, and bit him. If Michael gets the idea that Koko is being hurt, he will bark and bang into the door dividing him from his distressed playmate.

19

After School Gets Out

Like any student, Koko does not spend her day entirely in scholarly pursuits. Besides her periodic roughhousing with Michael, she likes to tumble around with those humans who are willing to play. During her early years one of her favorite games with me was what she dubbed a *spin*. After asking, Koko would lie on the floor on her back while I obligingly spun her around by the foot or hand. There was so little friction between the hair on her back and the linoleum floor, and Koko was so adept at balancing her weight properly, it was not difficult for me to get her 150 pounds spinning at terrific velocity. Afterward Koko would lie happily dazed and slightly dizzy on the hard floor. (It's not easy to get her really dizzy.) At Stanford there was a small jungle gym and swing outside the trailer, and Koko enjoyed a variation of the spinning game in the swing. With Koko sitting in the chair, we'd wind up the ropes of the swing like the rubber band on a paper airplane and let her go. Koko endured this torture with an expression of sublime happiness.

Koko has very pronounced likes and dislikes when it comes to the question of whose company she prefers. She tends to like quiet people. She does not like squealing children and loud people. She does like babies, and when she was young she was very sweet with the newborn child of one of my early volunteers. On the other hand, she decidedly did not like the baby chimp we brought to visit on one occasion.

Koko does not like people who are verbally aggressive with me. She likes Jane Goodall and Dian Fossey, both of whom have vast experience dealing with apes, but she had no special rapport with David Premack or Duane Rumbaugh. She tends to be hostile toward teenage girls, and she was rude to Lily Tomlin, who visited with a noisy entourage of filmmakers.

Jane Goodall once used Koko and Michael as expert witnesses when making a point to her associates at Gombe Stream Reserve, where she has been studying wild chimp behavior for many years. She wanted to stress that wild chimps were more comfortable in the presence of people who were crouched or sitting than with people who were standing. She asked us to ask Koko which she preferred, writing: "I'd love to be able to tell the Gombe chaps 'straight from the horse's mouth,' as it were." On several occasions we asked Koko, "Do you like people to stand up or sit down when they're with you?" She clearly and emphatically responded *Down!* with both hands to each inquiry but one, when she stressed her point even more by prostrating herself.

The trailer is not well equipped for visitors, and so we cannot have many come by. Consequently, the arrival of a visitor is something of a rare treat for Koko, and she will importune me to let the visitor in.

"Want friend in?"

In ask good, Koko replies.

Or when she sees visitors standing outside the door: *Do out out visit there.* When they come in: *Happy good you come.*

Not so long ago, Drs. Alan Skolnikoff and Suzanne Chevalier-Skolnikoff came by to visit. They arrived about 5:15, the time we usually start Koko's bedtime routine such as putting up the tarps on her windows and cleaning up her room. Suzanne brought a present for Koko, a bird puppet that she showed to Koko from the porch outside the trailer.

Koko, seeing Suzanne, signs *come.* Koko and Suzanne converse a bit about the puppet. Suzanne asks Koko "What's this?" while pointing to the bird puppet's tongue. Koko at first does not respond, but then signs *That red red* and points to the puppet's mouth. After a moment inspecting Suzanne, Koko signs *Time you* to Alan Skolnikoff. Alan comes over to the chain link and Koko investigates him briefly. Koko then returns her attention to the puppet. *That tongue... tongue there*, she signs, pursuing the earlier line of the conversation. At this point Suzanne and I begin talking for a few minutes. Just before 5:30, the time I ordinarily put up the drapes, Koko signs *Eager drapes do hurry*. The conversation continues in a desultory way until just after 5:30, when it is time for the Skolnikoffs to leave. As I open the door for them to leave, the Skolnikoffs say goodbye. *Good bye*, signs Koko. The Skolnikoffs linger at the door for a few minutes.

Visit Koko good, signs Koko, perhaps as a gentle prompt. The Skolnikoffs still have not left.

Look good go, signs Koko. I often use *look* as an imperative in English when I am impatient.

Bye, signs Koko in more polite form, but the Skolnikoffs are still chatting at the door.

Penny open ask go, signs Koko. The Skolnikoffs ask what Koko has signed and then leave.

The incident is revealing not only in what it suggests about the protocols Koko sets for visitors, but also because of the comment it offers on the claim that apes do not "perform" for strangers, because strangers do not know the "cues" that are responsible for eliciting the ape's signing behavior. Koko does sign with visitors once she overcomes initial shyness.

As Koko has grown larger, some of her favorite games—such as riding around on shoulders—are out of the question. Koko accepts our limitations with good grace. In any event she is never as rough with us as she is with Michael. Some of her male visitors, like Eugene Linden, a former wrestler, have no trepidation about tumbling around with Koko. Koko calls Eugene *Arm* and usually greets him with a request for a *chase tickle*, a game somewhat like a combination of hide-and-seek and tag—except that, because there are few places to hide in the trailer, it requires the human to act surprised a good deal. The game usually ends with a tussle.

With most visitors, however, Koko is extremely gentle. She extends her hand, leads them around her room, sits down with them, and puts her face close to theirs. She likes to comment on their clothing and ask them to eat something, or to give her something to eat.

Koko has a number of toys—balls, a skateboard, a variety of plastic playthings, and a motorcycle tire. She particularly enjoys it when someone sneaks up on her with the motorcycle tire and then shoves it over her head and shoulders.

In the excitement of a tussle Koko will sometimes mouth or scratch her playmate. Considering the power of her jaws, these infractions are mild, and never break skin.

I think that the scratches are intentional (maybe that's why she signs *Nails nails nails* as a threat), although Koko does not intend them to hurt anyone. Rather, they seem to be expressions of Koko's innate drive to establish dominance, a reminder to her human friends that although she enjoys their company, she is a gorilla.

On warm days I often take Koko out for walks or drives. Until our recent move from the Stanford campus we'd generally go to the lawn and park that surround the University Museum, which is across from the lab animal enclosure. We took these walks on Sunday mornings or during the summer when there were not many people about. We discouraged people from coming too close to Koko, and most of the students around Stanford seemed to understand and watched her antics from a respectful distance. On occasion she would pull me over to the administrative entrance of the museum and open the door. Considering that a gorilla had just come storming into their domain, the curators and secretaries who worked there took Koko's impromptu visits quite well. On the other hand, Koko, with her gorilla love of noise and bluster, tended to slam the door on her way out.

One of Koko's outdoor games is keep-away. But in Koko's rules, Koko always gets the ball. When one of Koko's deaf volunteers grabbed the ball and ran off, Koko—who was then four—hied after him and nipped him on the rear. I gave Koko a severe scolding and then fifteen minutes later asked her why she'd bit the man. *Him ball bad,* was her explanation.

On another occasion when Koko was five, she was playing a game of chase with Eugene in front of the museum. In her excitement, Koko gave Eugene a small bite when he caught her.

"What did you do?" I demanded instantly.

Not teeth, was her innocent explanation.

"Koko, you lied!" I replied.

Bad again Koko bad again, admitted a contrite Koko.

Lying, of course, is one of those behaviors that shows the power of language. Structurally it is similar to certain types of humor in that it exploits the conventions of language to convey a false reality. A lie is one of the most dramatic examples of propositional behavior, and, like humor, it is also one of the most difficult to prove because it is subjective. But, as is the case with a number of other elusive aspects of language that Koko uses, there is persuasive evidence that her lies are not mistakes but intentional efforts to avoid punishment for various misdemeanors.

It is not that Koko is badly behaved; after all, it's not her fault that most of the objects in her life have been designed for use by us feeble humans. Inevitably things get broken. Imagine how we would fare in a room whose furnishings were made of balsa wood, which offers about the same resistance to the average human that our accoutrements offer to a robust gorilla. Actually Koko is remarkably deft and controlled in dealing with the delicate artifacts of human life. Still, she does break things—such as the sink of her trailer.

When Koko was five she sat on the sink in the kitchen, causing it to cave slightly away from the counter. I then had an assistant named Kate Mann, a deaf woman Koko loved to blame things on. Kate was not, so far as I had noticed, prone to fits of violence. Thus, when I saw the broken sink and asked, "What happened here?" I was not disposed to believe Koko when she signed, *Kate there bad.* In fact, the image of a gorilla accusing a good-natured young woman of breaking the sink made it hard for me to maintain my pretense of irritation.

Koko's first apparent lie also involved blaming some destruction on Kate. In May 1974, at age three, Koko was asked, "Who broke this [toy] cat?" Koko replied, *Kate cat.* It was not entirely clear that this was a lie, because Koko had been with Kate the day before when the toy cat was broken, and she may have been summoning random associations. However, in some of her more recent lies, her intent is so clear as to be comical.

One such recent incident I recorded on film in 1978. While I was occupied with Koko's checklist of signs, I noticed that she had snatched a crayon from the top of the videotape monitor and was chewing on it. "You're not eating that crayon, are you?" I demanded. Koko signed *lip* and began moving the crayon first across her upper then her lower lip as if applying lipstick. Even Koko realized that this explanation was absurd, and when I asked her what she was really doing she signed *bite*. When I asked why, she responded *hungry*.

Koko tried a similar deception once when she was caught in the act of trying to break a window screen with a chopstick she'd stolen from the silverware drawer. When asked what she was doing, Koko replied, *Smoke mouth*, and placed the stick in her mouth as if she were smoking a cigarette.

Koko has used one accoutrement of the trailer to create inadvertent havoc. She loves to talk on the telephone. Rather, she loves to listen on the phone and attempt to talk. About the best she can do in "saying" words is to make a huffing sound, which can become very loud and hoarse. On the phone Koko sounds like a heart-attack victim or what in the lexicon of obscene telephone calls is termed a "breather." She also likes to make kissing sounds into the phone. The problem occurs when Koko calls "out." Occasionally, when I am otherwise occupied, Koko picks up the phone and dials numbers, and from time to time she gets a legitimate university number. Once I noticed that she had reached someone who did not hang up in the face of her huffing and puffing. When I took the phone and explained to Koko's telephone mate that he or she had been talking with a gorilla, there was a long pause and finally a disconnecting click.

Another time that Koko reached a university number, Ann Southcombe and I were talking and failed to notice that Koko was up to her old tricks until we heard a woman's voice say, "Don't hang up." The next thing we heard was Koko being connected with an outside operator, and the woman saying, "I think someone is choking." It was a couple of minutes before Ann and I could stop laughing long enough to call off the emergency.

One of Koko's favorite after-school activities is the car ride. When we were at Stanford, we'd usually stop at one of the campus soda machines and allow Koko to buy herself a drink. Not surprisingly, Koko committed to memory the various routes that led to the soda machines salted throughout the campus. One sunny July day when Eugene was visiting, we all piled into my aging Datsun for a ride, Koko grunting with pleasure at the prospect. Ron drove, Koko and I perched in the front seat, and Eugene sat in the back. As usual I asked Koko where she wanted to go, and as usual Koko pointed in the direction of the nearest soda machine, in this case at the rear of the Anatomy Building. Koko gazed longingly as Ron drove by the machine, and I asked her which direction we should turn next. Koko directed us around the Mausoleum and then pointed us in the direction of Ventura Hall, where there is another soda machine. Along the way we passed a few joggers and bicyclists, most of whom did not even give Koko a second glance.

Ventura Hall houses the Institute for Mathematical Study in the Social Sciences, headed by Patrick Suppes. Because Dr. Suppes has been involved in Project Koko, the logicians and programmers who inhabit the graceful old building were not surprised to see a gorilla walk up to the soda machine in the rear entrance hallway and put coins in it to buy a drink. With the soda in her hand, Koko walked happily back to the Datsun. Koko knew that we would now head back to the trailer, and so, when I asked her which direction we should take from Ventura, she pointed in the direction dead opposite to the way home.

Now and again there were Sunday outings to the Djerassi Ranch. Carl Djerassi has a large ranch in Woodside which his son Dale invited Koko to visit so that she might have some opportunity to wander about in a spacious wooded setting. This seemed to us a fine idea. On the other hand, Koko at first was none too enthralled with the great outdoors. On our first visit in 1975, Koko asked to have her leash put back on, and she wanted most of all to be inside the house.

But Koko quickly grew to enjoy the outings at the Djerassi ranch. As soon as we'd arrive, she and Michael would head for the trees. Both love to sit in the branches, clapping and slapping their chests in excitement, nibbling on leaves and bark, and eating sap. (Persimmon is by far their favorite tree to eat.) When not lolling in the trees or sliding down their trunks, Koko and Michael enjoy dangling from the branches by their arms and inching out toward the end, until the branches snap off and both gorillas tumble down in a shower of twigs and leaves.

The sight of Koko and Michael romping freely in these green hills first gave me the idea of finding a place where they might be like this every day, and not just on a rare Sunday outing.

20

The Gorilla View of Things

With all the concern about satisfying the requirements of scientific rigor, one promise of the experiments tends to get lost. That promise is insight into the gorilla view of the world. I have created a special file for conversations designed to elicit such information. The following conversations are all taken from what I call the "unique file," which consists of innovations and intriguing responses Koko makes in circumstances other than the formal controlled testing of her language skills. As the following conversations show, either I or one of my associates will introduce a topic of conversation, often with the hope of getting Koko to reveal her feelings and thoughts about various things. For instance, Barbara Hiller has repeatedly asked Koko to teach her gorilla talk. It is evident when Koko is not paying attention or is being purposefully obtuse, or simply does not understand what is being asked of her, but Koko can also be startlingly clear in her replies. Here, for instance, Barbara asks Koko about how gorillas feel about certain things:

BARBARA: Okay, can *you tell-me how gorillas talk?*
(Koko beats her chest.)
BARBARA: *What* do *gorillas say when* they're *happy?*
KOKO: *Gorilla hug.*
BARBARA: *What* do *gorillas say to* their *babies?* (Koko beats her chest.)

BARBARA (a little surprised): *What do you say to your baby?* (meaning her favorite doll, which she calls her baby)
KOKO: *Love Koko...gimme nut.*
(Barbara hands her a nut.)
BARBARA: *What do you say to Mike when you play?*
KOKO: *Mike Koko love.*
BARBARA: *What scares gorillas?*
KOKO: *Hat...dog.*
BARBARA: *Hats* and *dogs* scare gorillas?
KOKO: *Gorilla.*
BARBARA: *What do gorillas say when tired?*
KOKO: *Gorilla sleep.*
BARBARA: *What do gorillas think is funny?*
KOKO: *Clown...bug.*
BARBARA: *Enough?*
KOKO: *Love Koko...Koko love.*
BARBARA: *What?*
KOKO: *Drink apple.*
(The reader might recall this theme from earlier conversations.)
BARBARA: How about a *grape drink?*
KOKO: *Koko love drink.*
BARBARA: *You're very smart.* Are *all gorillas smart?*
(No response.)
BARBARA: Are *gorillas smart* or *stupid?*
KOKO: *Smart Koko.*
BARBARA: *What do you hate?*
KOKO: *Love.*
BARBARA: That's *nice, you love everything?*
KOKO: *Candy...fruit.*

At first glance certain parts of this conversation might give the impression that Koko does not know what she is talking about. But there are consistencies that run through numerous conversations on the same subject. For instance, beating her chest is Koko's idea of how gorillas talk, and a quite natural one at that, since that is something she and Michael do, but people don't. It might seem odd that Koko would mention hats and dogs when asked what she was afraid of, but these answers referred to two incidents that occurred just before this conversation. In the first, we were walking in the little park area across from Koko's trailer at Stanford when two elderly women with enormous sun hats approached. They came too close for Koko's comfort and did not seem to understand that the presence of two strikingly dressed strangers might be upsetting to Koko. Koko sensed my uneasiness and must have retained bad memories of the incident. The second occasion terrified not only Koko but me. Again we were walking, when an enormous dog charged. I dashed between Koko and the dog, shouted, and stamped my foot to intimidate him. He ran off when his owner called him away. Thus Koko's answers to what scares gorillas may not have been quite as random as they appeared.

Sometimes Koko seems not to understand a question, but then it will dawn on her what we are asking. For instance, on November 9, 1978, Maureen Sheehan, who has a gift for asking Koko good questions, had to search around before she came up with a form Koko understood.

MAUREEN: Do you *think all people know sign* language?
(No response.)
MAUREEN: Does *every person know sign* language?
(Still no response.)
MAUREEN: *Maybe* that's *too hard—who knows sign* language?
KOKO: *Koko.*
MAUREEN: *Right. Who else?*
KOKO: *Marjie.*
MAUREEN: *Yes. What about Mike?*
KOKO: *Maureen good.*

Another time when asked what language she spoke, Koko responded, *Sign me.*

Maureen has also extensively questioned Koko about the differences between gorillas and people. After talking about the differences between a girl and boy doll, Maureen turned to the subject of her and Koko.

MAUREEN: *What*'s the *difference between you* and *me?*
KOKO: *Head.*
MAUREEN: And *how* are our *heads different?*
(Koko beats on her head with her open hands quite hard, harder than a person would ever do.)

MAUREEN: *What else* is *different* between us?

(Koko does a gesture on her stomach resembling *blanket*.)

MAUREEN: Do *you mean something about stomach?*

KOKO: *Stomach good that.*

MAUREEN: Oh, but *what* were *you* saying *about blanket, different?*

(Koko moves her two hands up and down her torso, then pulls
    the hair on her belly.)

MAUREEN: *Now* can you *name something* the *same?*

KOKO: *Eye.*

MAUREEN: *Yes*, that's *right*, we *both have eyes.*

KOKO: *Love.*

Some days Koko is not in the mood to converse at all, and, like
a bored child, needs entertainment. One afternoon Barbara
Hiller found Koko looking at a catalogue, seeming listless and
bored. Barbara suggested that she bring it over so that they
could look at some pictures together. Koko ambled over with
her catalogue and sat down in front of Barbara. Attempting to
stimulate a discussion, Barbara started by asking her to find an
interesting picture. Koko turned a few pages and then pointed to
a photograph of a very dull pot. *That*, she signed.

"That's interesting?" asked Barbara.

That, Koko reiterated.

"Well, how about showing me a pretty picture?" Barbara
asked. Koko again flipped a few pages, and then pointed to a
nondescript roasting pan. *That*, she pointed.

"You really think that's pretty?" Barbara responded.

That.

Barbara, persistent, continued, "Well, okay, will you find a
scary picture?" Koko put down the catalogue, picked up her big
mirror, and pushed it right in front of Barbara's face.

We have asked Koko to delve into her vocabulary to come up
with definitions, opposites, and free associations. Asked "What
is an orange?" Koko replied, *Food drink*. Asked what does
"wrong" mean, Koko signed, *Fake*. Asked what is the opposite
of "first," Koko replied, *Last*. Another question: "What is a
problem?" Answer: *Work*. Question: "What is darkness?"
Answer: *Trouble know drapes*. When I asked her what is a
stove, she took the easy way out, simply pointing to the
microwave oven. I continued, "What do you do with it?" She

answered, *Cook with.* Perhaps the most revealing of these exercises are free associations, in which we ask Koko "What do you think of when I say _____?" Here are a few examples from May 30, 1980:

"What do you think of when I say drink?"
Sip.
"What do you think of when I say Lyna [an assistant]?"
Girl nut lip.
"What do you think of when I say shoe?"
Foot.
"What do you think of when I say Koko?"
Me.
"What do you think of when I say wash?"
Good clean.
"What do you think of when I say Mike?"
Toilet devil.

One of the most intriguing words in Koko's vocabulary is *drapes,* for which she has assigned many different meanings. The particular drapes in question are tarps used to screen out light and sounds from Koko's room when it is time for her to go to bed. Anytime something outside is bothering Koko, she will ask to have the drapes put up. On October 30, 1978, there had been trucks outside the trailer since morning:

KOKO: *Time drapes do.*
MAUREEN: *Why?*
KOKO: *Nice drapes close.*
MAUREEN: *Why,* is there *something scary?*
KOKO: *Hurry you drapes do.*
MAUREEN: *Why* do you *want* the *drapes?*
KOKO: *Close drapes... blanket drapes do hurry.*
MAUREEN: *Why,* are *you afraid?*
KOKO: *Love drapes drapes.* (Koko's *drapes* signs are getting more insistent.) *Drapes do good.*

Clearly, Koko can be very persistent and single-minded, and language offers a means of conveying urgency. Earlier, on July 12, Koko had also wanted the obscure security afforded by the drapes:

MAUREEN: *Why* are *you afraid—are you nervous?*
KOKO: *Nervous drapes me.*
(Maureen goes out to put up the drapes, and Koko starts banging around.)
MAUREEN: I *don't* put up the *drapes if you* are *bad.*
KOKO: *Nice gentle me go good.* (She kisses her own hand.) *Drapes hurry.*

Koko has confirmed her feelings about the comforting presence of drapes in other conversations. When Barbara Hiller asked Koko, "What do gorillas say when they are scared?" Koko replied, *Hurry drapes.* But the word *drapes* has meanings beyond security for Koko. Lately, Maureen and I have been asking Koko about her feelings about death. Koko has seen dead animals on the road, and suffered the loss of her pet goldfish, tadpole, and toads. Besides this, she has inflicted death on many insects. The feelings she expresses about death are all the more interesting because so far as we know subjects like burial have never been discussed in any detail in her presence. On December 8, 1978, Maureen asked Koko to pick the gorilla skeleton from among pictures of four types of animal skeletons. After Koko picked the correct skeleton, Maureen asked Koko whether the gorilla was alive or dead.

KOKO: *Dead drapes.*
MAUREEN: Let's *make sure,* is *this gorilla alive or dead?*
KOKO: *Dead good bye.*
MAUREEN: *How* do *gorillas feel when* they *die—happy, sad, afraid?*
KOKO: *Sleep.*

Koko seems to think of death as peaceful and secure. Several times she has used the word *drapes* to modify death. This impression is reinforced by the way she links the feeling of death with sleep. There is also evidence that this is not a case of mere confusion, since Koko gets quite upset when asked what will happen when she or I dies. Once when Maureen asked, "Do you think Penny will die?" Koko fidgeted for about ten seconds and then only signed, *Damn!* On the other hand, if the talk is about death in general Koko does not find the subject so terrifying:

MAUREEN: *Where* do *gorillas go when* they *die?*
KOKO: *Comfortable hole bye.*
MAUREEN: *When* do *gorillas die?*
KOKO: *Trouble old.*

We do not know where Koko got the idea that the dead go to a hole, unless it was from leafing through magazines (she is an avid "reader" of *National Geographic*). She has occasionally said *hole*, when asked where one goes when dead. Once, asked what makes her nervous, she said *Stop hole.* She said this before we knew of her association of holes and death. But again, her frequent use of the word *comfortable* reinforces the impression given by her use of the word *drapes*—that death is peaceful and secure. Finally, although she may see death as peaceful, she seems to realize that creatures die when in trouble or old.

Koko takes death quite literally. Once while she was idly playing in the room I was complaining to Maureen about the rigors of giving lectures. At one point I said, "I can't go to L.A. every month—it would kill me." I looked over at Koko, and saw that she was signing *frown.*

If Koko is already developing notions about the end of things, she has also given evidence of some grasp of the beginning of things, namely birth and related matters. For instance, consider this conversation of October 20, 1978:

MAUREEN: *Where* do *gorilla babies come from?*
KOKO: *Koko.*
MAUREEN: *Where in Koko* do *babies come from?*
(Koko points to her abdomen.)
MAUREEN: *You* are a *smart gorilla.*

Koko has seen pregnant women. Lee White, a former volunteer, visited while pregnant, and we told her that Lee was going to have a baby. After the baby was born, Lee returned with the infant, and we explained to Koko that this was the baby we had told her Lee was going to have. So it is possible that Koko put two and two together. On the other hand, it is also possible that Maureen unconsciously cued Koko by looking at her stomach. I was curious to know whether Koko understood birth, and on January 19, 1979, I asked Koko the same question: "Where do gorillas come from? Where are gorillas born?" Koko's reply was *up*, and she touched the ceiling. Perhaps she has reverted to believing in the stork, or I may have been looking at the ceiling. We will have to question Koko more about birth before we can assume any precocity.

What comes through in these conversations is Koko's ability to build complex ideas through a series of short "sentences"— just the open, creative use of language that is expressed in English through word order. What also comes through these conversations is the sense that Koko exploits different associations of the same word to develop different meanings. This capacity to use the metaphoric potentials of her vocabulary gives Koko the ability to express more sophisticated thoughts than the relatively short nature of her "sentences" might suggest.

21

The Limits to Koko's Learning

For all the expressive and creative ways in which Koko uses sign language, she clearly is not the peer of a normal speaking human. Just as clearly, there are differences between her use of sign language and that of the average deaf person. Understandably, though, scientists are reluctant about claiming what it is that separates human language from the potentials of animal communication, or even whether there are intrinsic differences between the two. The education of Koko and her chimp colleagues has also been an education for the behavioral sciences. Before these experiments, most scientists did not believe that any animal was capable of symbolic thought. A few decades ago there was active debate about whether any other animal besides man could be said even to be conscious. Now the debate is whether apes are capable of understanding and generating complex grammatical structures. Testimony to the turmoil produced by teaching language to apes is that these experiments have raised questions about whether a number of grammatical conventions such as word order are central to language, or specific to the linear, sequential nature of spoken language. Koko's creative uses of sign language suggest that there are other ways besides word order to make a language productive. At the least, Koko and her signing chimp cousins have stripped language of some of the mythic baggage that hindered comparisons with animal communication. As Mary Midgely, author of *Beast and Man*, has noted, "To claim that an elephant has a trunk is not to say that no other animal has a nose."

But what is it that distinguishes Koko from the average human? It is unsound to assert that an animal cannot do something just because it has not yet done it. When it became clear that apes were capable of understanding and using words, a number of critics tried to prove apes did not have "language" by producing lists of behavior—such as lying, asking questions, or joking—that a chimp or gorilla had not yet done. As we know, these criticisms were premature. Other criticisms of areas in which the evidence is inconclusive—such as word order—involve controversial criteria.

Indeed, there is very little that can be said about language today that is not open to question or controversy. Moreover, there is an enormous amount of information about ape language use that has not been assimilated by the various disciplines concerned with language and comparative psychology, and there is an enormous amount of raw data on Koko that has not yet been analyzed. Koko may have a number of surprises for the behavioral sciences not just in her future performance but also in this mass of raw data from her past. Until I have complete confidence that I have determined the limits of Koko's abilities, and until linguistics arrives at a consensus on what language is, it would be imprudent to assert that any particular ability distinguishes Koko qualitatively from the average human.

However, if we return to the case-study approach to Koko's use of language, it might be possible to determine whether there is a general and broad constraint that limits the use and development of her intellectual abilities. In comparing man and gorilla, it is important to keep in mind that it is unfair to measure one animal against the highest level of achievement of another animal. Nor should one assume that the highest achievements of a society are equivalent to the goals of evolution. Evolution is not goal-oriented. The fact that man can fly to the moon does not necessarily mean that flying to the moon is what makes us human.

Rather, we have to start with the common-sense assumption that gorillas are no more trying and failing to be human beings than we are trying and failing to be gorillas. It is unwarranted to assume that gorilla evolution is heading in our direction but is just a little slow in getting there. Rather, each species represents a different solution to the problems of survival. Our solution seems to depend heavily on the ability to respond flexibly to environmental challenges. The rat and the cockroach are also quite flexible in response to environmental changes. The difference between their flexibility and ours is that we possess greatly enhanced abilities to manipulate the world around us to ensure our safety and a regular food supply. Rather than be prey to the vicissitudes of nature, we decided to take matters into our own hands and organize our foodgathering to make the outcome more predictable. This involved cooperative hunting, the use of snares and weapons, and, ultimately, farming. This interventionist approach has had its radical effects on human evolution. It demanded a brain specialized for language and propositional thought, and, as importantly, access to this propositional mode. This is no small thing.

Inevitably, the path taken in human evolution involved selection for a propositional response to environmental challenges. It had to, because the enhancement of such abilities occurs at the expense of physical capabilities. Blood that previously invigorated the muscles increasingly had'to go to the brain. As neurologists have noted, metabolically the brain is an expensive piece of equipment. A bigger brain meant weaker muscles. A chimp roughly the same size as a human has a brain two-thirds our size and muscles three to five times stronger. This is not to say the chimp is that much dumber than man; rather it is to say that nature is efficient and thinking requires a trade-off of purely muscular vitality.

What is intriguing about the differences between man and ape is the question of *access* to whatever propositional abilities the species possesses. In order to ponder some problem, it is necessary to have a psychic space, a blackboard upon which to work, *displaced* from the pressing demands of the moment: One must be able to ignore immediate circumstances in order to retreat into the abstract, symbolic world of propositional thought. This requires, among other things, anatomical interconnections in the brain that allow sense information to be pooled without being acted on. If every piece of sense information were interpreted as an instruction to act, one would not have time to think. The surface evidence suggests that man is better equipped to retreat to this world, if for no other reason than that we do so more often.

A human is much more comfortable using language than is a chimp or gorilla. Eugene Linden, who has observed both sign-language-using chimps and Koko, notes that both Koko and the chimps often terminate a sign utterance with some action or other. A large percentage of Koko's sentences are directed toward achieving some object or requesting some action, like being tickled or watering the plants. Koko will sign to herself during play, and as she has gotten older her signing has become much less self-centered, but it is fair to say that she does not enjoy idle chat to the degree that we humans do. Eugene notes, though, that Koko does seem to resort to language more often than do those chimps he has observed.

One obvious advantage we have over the apes is that, equipped for speech, we can talk while our hands are otherwise occupied. If an ape is to use sign language, that desire must actively compete with the other activities for which ape hands were designed. Quite simply then, one reason that chimp and gorilla sentences remain relatively short even as their vocabularies increase is that there is a premium on the amount of time that the hands may be given over to activities like communication. However, more importantly, the shifts between sign language and other activities suggest that the propositional mode exists in tension with that immediate, intensely physical mode in which Koko spends much of her day. The gorilla is evolutionarily committed to a more visceral response to the world than we are. It has not mastered its emotions to the degree that we have.

One thing this suggests is that there is a lot more going on in Koko's mind than she expresses through sign language. In some respects she is like a stroke victim whose language production is damaged while comprehension remains intact—uttering a simple statement is far more arduous for her than it is for us. It may be still more arduous for the chimpanzees.

If there is a premium on economy of expression because of competing authorities seeking to govern the use of Koko's hands, it would help to explain why she chose to make her sophisticated modifications of Ameslan in the manners described. She can play with the structure of the sign language, inflect signs, and otherwise add overtones to utterances without adding length; such modifications add productivity to sign language in the way that relative clauses add complexity to spoken language. That Koko is not as loquacious as her human companions makes more sense if she is under constraint to pack as much meaning into as short an utterance as possible. Spoken language, on the other hand, both stemmed from anatomical changes and produced further anatomical changes. On the one hand, it arose from changes that permit us to deal with information detached from the pressing demands of the moment. At the same time it produced its own selective pressures —because it is so time-consuming—to increase our capacity to detach ourselves from the moment.

Nowhere do we mean to imply that the thoughts Koko is trying to communicate are as elaborate as the schemes the human mind can hatch. One difference already discussed is that Koko may lack some of the programmatic aspects of thought that are related to the linear nature of spoken language. But other considerations besides word order might prove useful as a comparative gauge of the development of Koko's conceptual abilities. One such aspect is the abstract sense of time.

It is important to keep in mind that the notion of time as fixed and cumulative is a Western idea, and one that has been modified by Einstein's theories of relativity. In other societies, time has different meanings. Polynesians typically take a subjective view of time; they see it as passing slowly during their childhood and much more quickly as adults. In New Guinea, natives extend time no more than five generations back—just beyond living memory. Ask three different generations of coastal natives when the world began and each will pick a time about five generations before theirs. However, given a Western education, all of these people are capable of understanding the various notions of time that occur in different cultures, and all are capable of learning the linguistic devices that permit discussing and understanding events either remote in the past or predicted for the future.

The utility of such devices is obvious. If one is to abstractly re-create an event or form a proposition, it is necessary to have symbols to mark the temporal relations of the events or ideas involved. How could one form a proposition without an understanding of these markers? Although Ameslan is less rich than English in precise specification of temporal relations, and although deaf children seem to have more difficulty learning temporal markers than hearing children, many researchers have concluded that the deaf have their own rules for the temporal relating of a sequence of events.

For instance, there is no formal expression of tenses in Ameslan as there is in English. Instead, what are described as "time indicator signs" modify a verb to specify when something happened. Deaf actor and Ameslan scholar Louis Fant describes the use of some these signs as follows:

> *Finish eat:* "I ate," "I did eat," "I have eaten," "I ate already," and so on.
> *Up till now eat:* "I have been eating."
> *Later eat:* "I'll eat later," "I'll eat after a while."
> *Not yet eat:* "I haven't eaten yet," "I didn't eat yet."
> *Will eat:* "I will eat."
> *Past eat:* "I ate" (sometime in the past).
> *Long time ago eat:* "I used to eat."

Fant writes, "Contrary to the popular notion that Ameslan has a weak and confusing time sense, it insists on a very specific denotation of when an act occurred, or will occur."

There is no doubt that spoken language, timebound as it is, is more elegantly equipped to specify the sequential nature of events. To some degree this reflects the constraints of a spoken language. But this does not mean that the deaf either do not understand or care about temporal relations.

When we are on a flight of abstraction, we enter a world furnished with symbols and a set of rules that relate to those symbols. The dimensions of this abstract universe are determined by a number of things. One major limitation is the amount of time any creature, whether human or animal, can afford to take off from the busy details of living to enter this world of hypotheses, analogies, and propositions. A physiological impatience to get back to the work of survival must limit the development of the abstract universe, and would also have its effects once the animal entered that world.

There is a difference between saying "War is hell!" and writing *War and Peace.* Part of that difference is that the work of

art specifies and otherwise makes explicit the myriad sentiments that give meaning to the phrase. For those GIs who have
said, "War is hell," those sentiments come from their immediate experience. If a great deal of one's reference is to immediate events, then Tolstoy's rhetorical devices are not necessary
to get across a point. The greater the remove in time of an event
and the greater the time span involved in the event, the more
elaborate the structures one needs to re-create that event.

In this sense we can see that the ability to specify fine
distinctions in temporal relations provides an indication of the
dimensions and elaborateness of this abstract universe. This is
simply a restatement of the importance of displacement.

Such an ability is only a crude yardstick by which to measure
an animal's intellectual and linguistic capacities. But it is easy
to see that without a sense of time as either a constant (as we see
it) or a variable (as both primitives and Albert Einstein saw it),
it would be difficult either to form an abstract proposition or to
describe an event that did not involve oneself. And if an animal
should prove capable of describing an intricate series of
abstract temporal relations, it would indicate that the animal
might, if nothing else again, be able to spin a good yarn. How
sophisticated, then, is Koko's sense of time?

There is no question that Koko can refer to past events. When I
brought up one of her misadventures that had occurred the
previous day, for instance, she signed, *Bed finished*, telling me
that she had already been scolded for breaking the bed. On
another occasion Koko persistently held up a toy spider ring and
signed *Ron*. Her companion kept asking Koko why she was
signing *Ron*. At first Koko signed, *That that that*, pointing to the
ring. Then she signed, *Funny*. The companion did not know
that Ron had worn the spider ring on his nose the week before
as a joke.

One morning in June 1980, as part of a test of Koko's
displacement ability, I pulled a plastic produce bag over my
head, cautioning Koko as I did so that this was a stupid thing to
do, something she should not do herself. The next day Ron asked
Koko what I had done with the bag. Her answer was, *Head
stupid*.

Koko has even used the word *yesterday*. On August 1, 1980, as
part of a conversation in which Koko had used the word *later*, I

said "What want later, to go out?" Koko responded, *Want yesterday.* Koko had in fact asked to go out the previous evening, but we had denied her request because she had misbehaved.

Koko, as one would expect, has her own internal clock, which is exceedingly well attuned to the regularities of her schedule. One day, shortly after she'd made the acquaintance of "Foot," the young construction worker she developed a crush on, she signed, *There look eager,* and pointed to the gate he was due to come through shortly.

Koko also has a sense of future events. If I promise to take her out the next afternoon, at the appointed time Koko will collect her sweater and leash in preparation for the event. But if I promise to do something "next week," the reference seems to be too vague for Koko to comprehend or remember.

The word *time* often appears in Koko's sentences as a command. Rather than referring to some abstract continuum, Koko uses the word concretely to mean "It is the moment for..."

Although she does not use them, Koko does understand the words *before* and *after*. As part of her evening routine with Barbara Hiller, Koko must clean her room before she is given a cookie. On July 24, 1980, Barbara asked Koko, "*What* do you *do to* your *room every night before* you *get* a *cookie?*" Koko responded, *Clean.* When Cindy Duggan asked Koko, "*Before I give-you* this *noodle, can you touch* your *head,*" Koko touched her head. Nor is she simply responding to requests in the order they are given, as she used to. When asked, "*Before you say ear, say eye,*" Koko answered, *eye, ear.* When I told her, "*Before you eat* your *banana, touch* your *tongue,*" she touched her tongue, signed *eat,* and ate the banana. Even when the request involves a more complicated action, Koko has responded appropriately: When I said to her, "*After* you *kiss* your *baby* [doll], *I'*ll *give-you* these *leaves,*" Koko turned, hunted for the doll and found it on the other side of the room, kissed it, and returned to me, hand outstretched for the leaves.

A good deal of what Koko signs is directed toward achieving an immediate practical result. Koko has no trouble understanding a relative clause if such understanding leads to some desired end. But if the road to the reward is sown with too many difficulties or if a complex task seems to lack purpose to Koko, she will lose interest in the problem, leaving it to us to determine whether her reluctance reflects mental limitations or laziness. In order to find out whether Koko is capable of articulating a complex series of relationships, I will have to find a problem that she wants to solve. In the meantime, what is revealing about her understanding of temporal relations is the self-interest and practicality that seem to govern her use of these words.

Koko has a memory of the events of her own life and can refer to them through sign language. She also has some understanding of the future, especially as it pertains to some desired goal. But even here, if Koko concludes that the reward isn't worth the work, she may ignore or give up on the problem. What I am suggesting is not that Koko is an egomaniac but that this apparent self-interest points to one constraint over her access to her propositional abilities and also may suggest a limit to the types of abstract structures Koko can employ.

Earlier we suggested that the propositional exists in tension with an animal's "natural" responses to the events of its day. When thinking about some novel problem, we suspend the urge to act and temporarily usurp control over our actions. Once the problem is successfully solved, the solution is stored so that it can be summoned without further thought. Thus, once Koko learns a sign, summoning that sign does not require the degree of access to her propositional mode that was required when she first made the associations necessary to learning its meaning. (When learning centers in human brains are destroyed, victims often can still perform a number of functions they learned earlier. They just cannot learn new skills.) Driving, for instance, is difficult for us to learn, but once learned it does not require much thought to operate a car. During those moments of learning or generating a novel utterance, Koko's "mind" is usurping control over her hands. Inherited behavior patterns have a relatively strong hold on a human's attention. We all know how hard it is to get a child to sit still. It is harder for a gorilla. And unlike the child, whom evolution has predisposed to explore the potentials of language, Koko uses language in a way that sometimes conflicts with other means of expressing herself.

Koko is in extraordinary circumstances, and language is critical to the human environment in which she is being raised.

But in the wild she could have gotten by without it, or at least with whatever vocabulary of gestures gorillas have developed to communicate with each other. We, on the other hand, are dependent on language for our survival.

Which is to say that the difference between Koko's linguistic abilities and ours is probably not attributable to one critical, magical factor, but rather to the chemistry of many faculties more developed in man than in gorilla.

22

Conclusions

Before Koko, Washoe, and other apes first showed they possessed some facility with symbols, people spoke with reasonable confidence about what it was that separated man from other animals. If there was disagreement about whether the difference was the ability to construct sentences, think symbolically, or create tools, at least there was broad agreement that there were intrinsic differences between human and animal intelligence. Debate centered on the question of what it was that made the human mind qualitatively different, and not whether it was qualitatively different at all. Language-using apes have eroded that earlier notion, and also exposed uncertainty over the proper definition of human intellectual abilities. The animal/human dichotomy that has guided our thinking about language has given the investigation of language a curious circularity. Starting with the assumption that there were no continuities between animal and human language, we have looked for evidence to support this assumption, and then used this evidence to justify the assumption.

This circularity lent the human intellect a spurious unity. By arrogantly ruling that there were no continuities linking animal and human thought, we fostered the idea that human abilities should be considered only in relation to other human abilities, not in relation to their animal correlatives. Testimony to this tendency is that while the general population is inclined to believe that human abilities are the product of divine intervention—or even, as some think, intervention from outer space—many are unwilling to accept the wonders of the intellect as the product of the development of abilities found to lesser degrees in other animals alive today. Carl Jung, among countless others, believed that man could not achieve self-knowledge through comparison with other animals. Thus, on the one hand we preclude valid comparison with the most likely candidates for shedding light upon our origins, and then complain about our alienation from the natural order of things. Nor is this a matter of simple prejudice. Behind Noam Chomsky's theory that there is an inherited deep structure of language one can see a creationist view of the universe.

Long before Darwin, there were scientists inclined to accept our ancestral connection to the natural order. Searches for the fabled "missing link" occupied theorists from Albertus Magnus onward. But even this idea that there must have been some intermediate creature—half-man, half-animal—implies that no real comparison can be made between man and animal until such a creature is discovered.

Darwin sowed the seeds of a perspective on man and nature that improved the climate for consideration of the communalities of man and animal. Evolutionary theory has changed markedly since he first hypothesized a common ancestor for man and ape, and indeed Darwin, who in many respects was quite stodgy and conservative, would probably be shocked at the implication of many of the experiments done in his name. However, for all the changes in the theory of evolution, it supplies a perspective that permits the search for significant continuities between man and animal. It is difficult to overstate how different the same behavior might seem when seen from an evolutionary perspective as opposed to the traditional animal/human dichotomy. When we look at man's higher intellectual abilities from an evolutionary perspective, we compare them with similar abilities in other animals, and the need to posit a "missing link" vanishes. The "missing link" between man and animal turns out to be a perspective that permits a revealing comparison of behavior across species.

Whether or not Koko and her chimp friends have thrown the scientific world into confusion depends on which perspective one takes on human behavior.

To clarify this point, consider some of the central criticisms of Koko's use of language.

Throughout this book we have looked at Koko in a two-tiered way, on the one hand assessing her performance according to criteria based on data collected in strictly controlled circumstances, and on the other hand attempting to interpret what Koko does with sign language of her own volition. These two approaches to Koko's use of sign language lead to radically varying estimates of her abilities. Innovations that must be scored as "errors" when they occur in a formal testing situation look entirely different when assessed in terms of the background against which those innovations occurred. Thomas Sebeok, probably the most vehement critic of the experiments today, ridicules the idea that Koko was making a joke when she executed the *drink* sign in her ear in the incident cited in Chapter 9. Sebeok claims that one can only conclude that Koko was in this case making a mistake. But is this a plausible conclusion?

No one disputes that Koko knows the sign *drink*. It's a sign that she has used appropriately many thousands of times. It is exceedingly unlikely that she had a memory lapse when Barbara insisted that she make the sign before giving her the drink that she wanted. Then there were the circumstances in which she made her error or joke. Koko, as is her wont, was being ornery that evening. Nobody disputes that gorillas can be cussed and stubborn. We also have corroborative examples of the many ways Koko uses sign language to express her willfulness. We have evidence that Koko knows how to adapt the structure of sign language to express her meanings—her "loud," exaggerated signs, her merged signs such as *Koko-love*, and her invented signs, to name but a few. Given all these contextual factors, it is stretching the point to assert that Koko forgot the word *drink*.

But there is a stronger argument that Koko knew what she was doing when she made the *drink* sign in her ear. Some critics of animal language experiments tend to view the subjects as purely passive, like pigeons in a Skinner box, dormant until stimulated to act by the experimenter. The animal is presumed to acquiesce docilely in the endeavor, content to receive its rewards after a hard day's work. For his or her part, the experimenter sits in the self-imposed bafflement of the double-blind setup (even here Sebeok sees the potential for cueing) and dumbly records the animal's responses, unaware of what it is responding to. This is fine during testing, but I did not simply shove Koko into a closet between testing situations. Rather, I would talk with her. The important part of the day for Koko, and the real "purpose" of learning the language for her, was the communication with her human companions. She is not a robot, but a sentient being, and we have equipped her with a means to make her intentions known to us. It is fair to say that Koko is more interested in the potentials of language in this regard than she is in getting scores of 90 on double-blind tests. In fact, as noted earlier, after Koko performed well on a test, her scores deteriorated when a test became too familiar to her. This is just the opposite of what should happen as she becomes accustomed to the conventions of a given test. Koko initially performs well in part because she enjoys novel activities and problems. But it is also possible that these initial good performances allow her to make a point about her boredom with routine testing which she would not be able to make if she failed every test. By first establishing that she can do well, Koko is then free to signal her displeasure on subsequent tests. She knows this frustrates me, and she knows the reason it frustrates me is that I know she can do better, and she knows I know that she can do better.

Similarly, when Koko was being ornery about making the *drink* sign, she expressed her displeasure in a way that gracefully underscored that she knew what she was doing, but was still refusing to do it right. She was aware that Barbara knew that she knew the sign for *drink*; Barbara has worked with Koko since the first days of the project. Had Koko randomly made any inappropriate sign, Barbara might not have understood her point.

Of course, making assumptions about Koko's intention is anathema to a true empiricist. We don't read meanings into the mistakes of a parrot: why should we make exception for a gorilla? The answer is that there is a mass of corroborative evidence that begs for consideration. One does not want an experimental design that hinders the investigation of what it is supposed to uncover. Moreover, if we read significance into the mistakes and overgeneralizations of children, why then should we deny out of hand the possibility that these same games might not occur in the learning process of what may be our closest nonhuman relative? It would be more suspicious if Koko did not play with her language, but only mechanistically answered the questions we put to her.

I may eventually be able to set up formal controls to gather data on Koko's sense of humor, but given that a few critics see the potential for cueing even in the double-blind testing of vocabulary, it is unlikely any experimental design could adequately prove to the suspicious Koko's intentions in these subjective activities. In the meantime, Koko continues to joke, and each new incident adds to the plausibility of this interpretation of her intentions.

According to one's perspective Koko is either a dolt who has only a shaky hold on a basic vocabulary, or a bright, playful, creative creature capable of quite sophisticated innovation. Nor is this difference merely a question of empirical method versus anecdotal material. Rather, underpinning the two perspectives are two conflicting views of man's place in nature.

By ruling as inadmissible the corroborative information of Koko's variant signs, critics are making a judgment about the nature of language. They are ignoring the communicative and interactive aspects of language. From an evolutionary perspective, language—whatever it is—is one of a number of means of communication employed by different creatures. When we humans communicate, we draw on a number of nonlinguistic cues to determine the meaning of a message, even if it is spoken in the most precise English. We attend to tone of voice, eye movement, gesture, and so forth, each of which comments on, and sometimes even contradicts, the literal meaning of a message. The rhythms of the speech stream set up expectations that we know what is to come before a speaker has finished a sentence. Language is a fugue in which the linguistic elements are but one voice. And yet for one school of linguists, syntax is the only crucial voice, obscuring all other elements of speech.

What syntax does is provide a set of conventions that permits a limited number of words to be combined in novel ways. It permits openness and productivity. But, as described in Chapter 12, there are other ways to make a language productive besides word order—Koko's inflections and modulations of signs, for instance. The legitimate question is not whether Koko uses proper word order, but whether she uses her acquired sign language in an open and productive way. I believe that Koko has demonstrated that she does so. And, if we look at the pressures for economy of expression (discussed in the previous chapter) that operate on her language use, it makes sense that she uses inflection and modulation rather than a string of relative clauses to complicate her "sentences."

To fully understand Koko's abilities in relationship to ours, we must look for differences in the light of what continuities there are between her use of sign and our use of language. If we stack the deck against the consideration of such continuities, not much is to be gained except some reinforcement of our supremacy in the natural hierarchy—a position that Koko has no interest in threatening. If, on the other hand, we take the opportunity that Koko proffers to demystify language, there is an enormous amount that Koko can tell us about the origin and nature of our abilities. And this is precisely the value of looking at Koko's performance with her intentions in mind. To do so allows us to integrate her linguistic behavior with other aspects of her social relationships with those around her. In considering Koko's intentions we should also consider those nonverbal abilities to make propositions that might be the source Koko draws upon and adapts in learning sign.

When we hamstring the comparison of human and ape abilities, we hinder our understanding of ourselves and our closest surviving relatives. We are in essence choosing not to know.

For my part, I find it more comfortable and nourishing to live in a world in which I can see and acknowledge elements of my behavior in the creatures around me, in which I can identify and communicate with a close relative with whom man has been out of touch for the past few million years. I find it refreshing that we have finally turned our gaze to the world around us, rather than looking to space to find someone quasi-human to talk with. It is not a moment too soon, either, because while we have been looking to the stars, most of our potential conversationalists on earth have been driven closer and closer to extinction.

In a sense, my career in psychology has followed the broad movement in the behavioral sciences. I have moved from the lab in which the animal seems to be valued only insofar as it can shed light on human problems, to the controlled study and testing of an animal's mastery of a human ability, to the appreciation of the insights into both animal and human behavior that follow when one attends to the animal's intentions rather than to our own. However, the key to this gradual broadening of my appreciation of animal individuality has been Koko's acquisition of human sign language, along with the rigorous testing and data gathering that first convinced me that Koko knew what she was doing when she used sign. I fully agree with those who demand strict controls before accepting that an animal has demonstrated some linguistic behavior. But, once collected, this hard data gives me norms against which I can check the less controllable uses of language that Koko demonstrates. This marriage of rigor and common-sense observation seems to be the most fruitful way to approach the study of our relationship to other animals. By common sense I mean that we should never lose sight of the fact that language is a means of communication.

My relationship with Koko has flowered during the course of Project Koko. It is possible that without our common language the bonds between Koko and me might still have developed, but on innumerable occasions sign language has allowed both Koko and me to express our feelings, to prevent misunderstandings, and to reassure ourselves of the other's affection and trust. A common language has permitted Koko and me to be explicit where without a common language, the relationship between human and animal might only be implicit.

Still, however close Koko and I have become, I have not lost sight of the scientific importance of her achievements. I still maintain the daily checklists and logs, the videotaping session, and other means of monitoring her performance. Indeed, Project Koko is the only one of the ape language-experiments in which there is a constant, uninterrupted record of progress. Koko is the only language-using ape who has received continuous instruction by the same teacher. She is the only language-using ape who has received nine years of language instruction. These circumstances give Koko added importance, especially in light of the new assaults that have been launched against the credibility of these experiments during the past two years. The suggestion by critics that imitation and prompting play a large role in ape conversation is based primarily on the performance of a chimp, Nim, who had many changes of environment and personnel during the four years of that experiment. I am currently analyzing videotapes and gathering other evidence to see whether Nim's was in fact a case of arrested development in language acquisition, a circumstance brought on by the environmental and methodological oddities of the experiment.

What is interesting about the new batch of critics is that some of the early pioneers in this comparative research have joined in. The Gardners, Roger Fouts, and I still believe that there is sound evidence to support the claim that apes use sign language spontaneously, appropriately, and creatively, while Herb Terrace and David Premack now have doubts about the performance of their respective pupils.

These are paradoxical times for the comparative study of language. Roger Fouts is studying the transfer of sign language from Washoe to her adopted infant Loulis. Loulis has already acquired a dozen signs from Washoe. I am continuing to collect data on the innovative use of sign by Koko and Michael. At the same time Herb Terrace writes that "'Chimp Language' appears to have the sole function of requesting rewards that can only be got by signing," and offers the problematical case of Nim as evidence that apes cannot master any more than the simplest sentences.

In considering this paradox, I cannot help but note that Terrace returned Nim to his owner in Oklahoma after four years, and that Premack's experiment with Sarah ended after a few years. Moreover, in none of these cases did the experimenter allow himself to develop a true, close rapport with his chimp. This was justified in the laudable name of objectivity, but given the sensitivity of the animals involved—Koko's signing is affected by even slight disruptions in her routine—it is hard not to wonder whether the different conclusions about ape language abilities reached by these scientists ultimately trace back to the different relationships between experimenters and subjects and to the persistence that has marked the efforts of those of us who have established close rapport with our subjects. If this is the case, I am reaffirmed in my belief that one cannot really understand the mental workings of other animals or bring them to the limits of their abilities unless one first has true rapport with them. Even the critics admit this possibility. What they fail to see is that the problem really is a misunderstanding of the purpose of language. Once that misunderstanding is straightened out and we accept language as a communicative behavior, the evidence of Koko's abilities is compelling for those who want to see it.

EPILOGUE

Koko, Michael, Ron, and I are now comfortably ensconced in Woodside, California. Getting there, though, was marked by no less drama than our earlier move from the San Francisco Zoo to Stanford. Today the lab animal enclosure looks empty to me, although I was never comfortable housing Koko in an area devoted to sheltering the future victims of medical experiments.

The decision to move to Woodside traces back to 1979, when my association with Stanford began to draw to a close. In April I was awarded my Ph.D. in Psychology based upon my work with Koko. Upon my graduation, Stanford stepped up pressure for me to move Koko and Michael. Officials were afraid—unjustifiably—that Koko might hurt someone and leave Stanford the defendant in a huge damage suit. I tried unsuccessfully to persuade different officials that this was extremely unlikely. I also tried to convince the university to lease to the Gorilla Foundation an abandoned half-million-dollar primate facility, which Jane Goodall had established at Stanford before it decided that it did not want to house any large primates on its property. Again I was unsuccessful. Then, as deadline after deadline passed for Koko and Michael's eviction, I began to search for either a position at another university, or a property that we might buy to house Koko and Michael.

After an exhaustive hunt, we discovered an old farm for sale in Woodside, about ten miles from Stanford. The farm is small—

212

only seven acres—but it is secluded and quiet, and has protected access by virtue of trees and the rough terrain of the area. The house is an old shingle dwelling without central heating. The property contains a large abandoned poultry house, which I immediately envisioned as a future home for Michael. The buildings are surrounded by fruit trees bearing persimmons, apples, pears, plums, apricots, and figs. A golden eagle nests in a pine tree a few yards from the house. The setting is lovely. It is high on the ridge line of the hills that divide San Francisco Bay from the Pacific Ocean, and from the living room it is possible to look down the rolling hills and redwood-studded valleys to the ocean.

The only problem was raising the money to buy it. Some funds were donated in response to an article on Koko in *National Geographic* and the reprinting of that article in *Reader's Digest,* and we also had a small income from fees paid for permission to use photos of Koko. Members of the Gorilla Foundation helped with additional contributions.* With these savings Ron and I put together the financing necessary to make a down payment on the purchase of this property. We took title to the property in September 1979, and moved Koko and Michael and the trailer up the hillside to their new eyrie on Halloween. We figured that a couple of gorillas riding in a car would blend into the background on that day. When we arrived, Koko and Michael hopped out of the car and started climbing the trees, sampling the fruit, and exploring the grounds and buildings.

As it turned out, a stowaway from the lab animal enclosure had also made the trip up the mountainside. During the first few days after the trailer was installed on its new site, I heard strange thumpings beneath the floor. Then two weeks later a workman reported spotting a rabbit under the trailer. It was the same fat black-and-white rabbit that had hung around the trailer when it was in the lab animal enclosure. Evidently he had made his home in the undercarriage, and decided to come along for the ride when we moved. I don't blame him. His future looks a lot sunnier now that he is far from the surgeon's knife. Koko's and Michael's future seems a lot sunnier now too. We are

---

*Readers who want further information about the Gorilla Foundation and its newsletter may write to 17820 Skyline Boulevard, Woodside, CA 94062 (or to P.O. Box 3002, Stanford, CA 94305).

not yet on an island where the gorillas might enjoy total
freedom, but at least we have found a satisfactory home, as long
as we can keep up with the expenses. If we ever do find an island
where Ron, Koko, Michael, and I and any gorilla offspring can
find refuge, I think I will invite the rabbit along as well. He's
made it this far.

BIBLIOGRAPHY

SELECTED ANNOTATED BIBLIOGRAPHY

Griffin, Donald R. *The Question of Animal Awareness: Evolutionary Continuity of Mental Experience.* Rockefeller University Press, 1976.
A scientist's view of the continuities that link animal and human consciousness.

Klima, Edward S., and Ursula Bellugi. *The Signs of Language.* Harvard University Press, 1979.
The most thorough study of sign language yet published.

Linden, Eugene. *Apes, Men, and Language.* Penguin, 1976.
A survey and interpretation of the various sign language experiments with apes.

Midgely, Mary. *Beast and Man: The Roots of Human Nature.* Cornell University Press, 1978.
A thoughtful interpretation of the philosophical questions raised by experiments exploring the intelligence and language of other animals.

Patterson, F.G. "Linguistic Capabilities of a Lowland Gorilla." Ph.D. dissertation, Stanford University, 1979 (available through University Microfilms International Edition).
A scientific consideration of Koko's language use.

Piaget, Jean. *The Language and Thought of the Child.* New American Library, 1955.
Piaget's design for the development of language in children.

Rumbaugh, Duane M., ed. *Language Learning by a Chimpanzee: The Lana Project.* Academic Press, 1977.
An account of Project Lana, in which the author taught the chimpanzee Lana to communicate through an invented "language" encoded on a computer console.

Schrier, Allan M., et al., eds. *Behavior of Nonhuman Primates*, Vol. 4. Academic Press, 1971.
Accounts of some of the pioneering experiments exploring the intelligence and language capacities of chimpanzees, including the Gardners' early work with Washoe.

Sebeok, Thomas, and Ann Umiker Sebeok. *Speaking of Apes*. Plenum Press, 1980.
An anthology of essays critical of the various language experiments with apes, assembled by the most vehement of the recent critics.

Terrace, Herbert S. *Nim: A Chimpanzee Who Learned Sign Language*. Knopf, 1979.
An account of the author's attempt to teach sign language to the chimpanzee Nim; despite the subtitle, Terrace about-faces in later chapters and asserts that Nim did not learn language.

PROJECT KOKO BIBLIOGRAPHY

Patterson, F.G. "The Gestures of a Gorilla: Language Acquisition in Another Pongid." *Brain and Language* 5 (1978): 72–97.

—— "Linguistic Capabilities of a Lowland Gorilla." In *Sign Language and Language Acquisition in Man and Ape: New Dimensions in Comparative Pedolinguistics*, edited by F.C. Peng. Boulder, Colorado: Westview Press, 1978. (Reprinted in Schiefelbusch, R.L., and Hollis, J.H., eds. *Language Intervention from Ape to Child*. Baltimore: University Park Press, 1979.)

—— "Conversations with a Gorilla." *National Geographic* 154 (October 1978): 438–65. (Condensed in *Reader's Digest* 114 [March 1979]: 81–86.)

—— "Human Communication with Gorillas." In *In the Spirit of Enterprise*, edited by G.B. Stone. San Francisco: W.H. Freeman & Co., 1978.

—— "Linguistic Capabilities of a Lowland Gorilla." Ph.D. dissertation, Stanford University, 1979 (available through University Microfilms International Edition).

—— "In Search of Man: Experiments in Primate Communication." *The Michigan Quarterly Review* 19 (Winter 1980): 95–114.

—— "Gorilla Talk: Comment on 'Monkey Business.'" *The New York Review of Books* 27 (October 1980): 45–46.

—— "Creative and Innovative Uses of Language by a Gorilla: A Case Study." In *Children's Language*, vol. 2, edited by K.E. Nelson. New York: Gardner Press, 1980.

—— "Can an Ape Create a Sentence? Some Affirmative Evidence." *Science* 211 (1981): 86–87.

Gorilla, a journal published biannually by the Gorilla Foundation, P.O. Box 3002, Stanford, CA 94305.

Koko, a Talking Gorilla, a ninety-minute 16 mm film by Barbet Schroeder. Available from New Yorker Films, 16 West 61 Street, New York, NY 10023.

Index

217